Automobile Batteries: A Practical Handbook
The Construction, Charging, Repair and Maintenance of Ignition, Starting, Lighting and Electric Vehicle Batteries: Dry, Lead and Alkaline Types

by Harold H.U. Cross

with an introduction by Roger Chambers

This work contains material that was originally published in 1919.

This publication was created and published for the public benefit, utilizing public funding and is within the Public Domain.

This edition is reprinted for educational purposes and in accordance with all applicable Federal Laws.

Introduction Copyright 2018 by Roger Chambers

Self Reliance Books

Get more historic titles on animal and stock breeding, gardening and old fashioned skills by visiting us at:

Introduction

I am pleased to present yet another title on Historical Science.

The work is in the Public Domain and is re-printed here in accordance with Federal Laws.

As with all reprinted books of this age that are intended to perfectly reproduce the original edition, considerable pains and effort had to be undertaken to correct fading and sometimes outright damage to existing proofs of this title. At times, this task is quite monumental, requiring an almost total "rebuilding" of some pages from digital proofs of multiple copies. Despite this, imperfections still sometimes exist in the final proof and may detract from the visual appearance of the text.

I hope you enjoy reading this book as much as I enjoyed making it available to readers again.

Roger Chambers

CONTENTS

PREFACE

THIS little volume is offered to the electrical and automobile fraternity, as a guide to the care of the modern automobile battery. Very little electrical knowledge has been taken for granted, so that the student, and the garage or service station mechanic who is taking up battery work, will not be perplexed either by imposing formulæ or recondite technical terms. Such terms as are unavoidable in a work of this nature are briefly explained in an appendix.

Although this work is purely *practical*, it has been thought wise to add here and there a little science of a most elementary character for the benefit of the many students who use these S. & C. volumes in pursuit of some college course which requires of them a knowledge of the practical aspects of electro-physics.

Grateful acknowledgments are here made to Messrs. Ward & Goldstone, the General Electric Co., Ltd., the Economic Electric Co., Richford & Co., and C. A. Vandervell & Co. for the loan of illustrations and material for this work.

PREFACE

The greater part of the text is based upon articles which have from time to time appeared in *The English Mechanic* and the *South African Motorist*, and has been made available by courtesy of their respective editors. Chapter II. was given as a lecture at a college in Los Angelos, California.

A word of special thanks is due to Superintendent Engineer Paterson and Chief Engineer A. C. Winter, of the Union Steam Ship Company of New Zealand, for acts of kindness shown to the Author upon many occasions.

H. H. U. C.

SAN FRANCISCO,
CAL., 1919.

THE CARE AND REPAIR

OF

AUTOMOBILE BATTERIES

CHAPTER I

THE "DRY" BATTERY

N.B.—For technical terms refer to Appendix, page 97.

THE modern motor-car is becoming more "electric" every day—in fact scarcely a week passes but some ingenious individual possessed of an inventive turn of mind makes the steps of the Patent Office red hot in his hurry to protect some new terminal, dynamo, lamp, battery, etc., for use upon the automobile. This ceaseless rush has been in progress for such an extended period that many ways and means have been employed to the end that the necessary electricity should be available for the efficient performance of these many labour-saving and delight-giving devices.

In the pre-magneto days, when every motorist had his coil and battery, ignition troubles were very frequent, as the principles that governed the successful operation of this battery were at first past finding out.

B

When, at length, the motorist had more or less mastered its funny little ways, the reliable and less troublesome magneto made its appearance and came to stay. Practical handbooks on Storage Batteries and how to make them store, lost much of their former appeal to the automobile fraternity. It was true that battery ignition was still used on many cars as a standby, and for easy starting on a cold morning; also where it was highly desirable to retain a high range of ignition flexibility, such as for racing purposes, or continued driving in heavy traffic. In the main, however, the battery was relegated to the limbo of things superseded.

Now that the days of car lighting and starting by electricity have come to pass, to say nothing of the ever increasing popularity of the many forms of electric vehicle, the battery has visited us again—an old friend in, perhaps, a slightly altered form. It is unfortunate that so long has elapsed between the disappearance and return of the battery, as the automobilist has almost forgotten its vagaries and idiosyncrasies, which, despite highly flavoured advertisements, have not been wholly eliminated. Manufacturers of storage batteries hold out great hope in connection with their cells—pointing out their excellent design, their super-solid construction, and the immaculate finish of their products. The real superiority of the modern storage battery is due to the improved conditions of service, as to-day a proper "charge" is a dead certainty, as in ninety-nine cases out of a hundred, charging gear in the shape of an automatically controlled dynamo is installed aboard the car, whilst to take care of the hundredth case a storage battery service station is situated in almost every street.

There are three distinct types of cells used in modern automobiles. Two out of the three are of the storage type—secondary batteries. The lead/acid cell, and the nickel-iron/alkaline cell. The third is the well-known but misnamed "Dry Battery" generally used as an emergency ignition device. In this present chapter we shall turn our attention to the salient points of the dry battery and show the reader how he may gain greater efficiency than is customary from this form of primary cell.

The so-called dry cell differs from the old-time Leclanché cell with its porous pot, glass jar and zinc rod,—still much used in Europe for working Bells, in the fact that it has certain features that render it of enormous commercial importance. Firstly, it is extremely portable as, as its name implies, there is no liquid to spill. Secondly, the internal resistance of the dry cell is so low that upon short circuit some 30 ampères can be drawn momentarily, and, furthermore, as it does not polarise so readily as the wet form of Leclanché cell, it can even be used in an emergency for lighting low candle-powered lamps. The limits of this present treatise will prevent more than the most cursory consideration of the physico-chemical aspects of dry cells; however, a few words may be of interest to the reader. The essentials of a Leclanché cell from a purely elemental point of view, are a zinc electrode, a carbon electrode which forms the positive terminal, manganese dioxide (MnO_2) as a depolariser, and finally a solution of sal-ammoniac (NH_4Cl) as the exciting fluid. A depolariser, it may be mentioned, is an agent which is exceedingly greedy for the nascent or new-born hydrogen liberated in the vicinity of the zinc electrode (negative pole). As this hydrogen, swept on by the current (which inside the cell flows from the zinc to the carbon plate), meets the manganese dioxide which surrounds the carbon pole, it is oxidised to water and within certain limitations the cell does not suffer reduction of output through the reverse current which would be produced if the hydrogen were to reach the carbon pole, and also the insulating effect of the hydrogen film which would thus be interposed in the internal circuit of the cell. It is the function of the manganese dioxide, then, to remove this unwanted hydrogen, which it does most effectually providing no large current is in requisition. If the influx of hydrogen due to excessive current is too great the manganese dioxide cannot act upon it and, in spite of it, the cell polarises. However, with this depolariser recovery is very rapid, and after a rest of a few moments the cell is as good as new. In the so-called dry form no porous pot is utilised, which lowers its internal resistance and enables more current to be taken from it. Recuperation does not appear to be so rapid, and for that matter neither is polarisation so rapid, as much more difficulty is experienced by the hydrogen ions* in their migration to the carbon element of the cell.

The chemical reactions of the dry cell are very complex in character, but in essentials they of course follow those of the wet type of Leclanché cell, which are as follows:—

$$Zn + 2NH_4Cl = ZnCl_2 + 2NH_3 + H_2$$

Zinc reacts with *Sal-ammoniac* to produce *Zinc chloride* and *Ammonia gas* and *Hydrogen.*

The above represents the main galvanic action, the hydrogen then drifts over to the carbon electrode surrounded by the dioxide of manganese, and is oxidised to water as follows:—

$$H_2 + 2MnO_2 = Mn_2O_3 + H_2O$$

Hydrogen the polarising agent } reacts with { *Manganese dioxide* the depolariser } and forms { *Manganese Sesquioxide,* a lower Oxide of Manganese } and *Water.*

Fig. 1 is a sectional view of an average dry cell; it will be noted that the principal parts are indicated by the letters: (a) is the cardboard insulating case which carries the zinc case or box (b); (c) is the paste composed of ammonium chloride and zinc chloride, water, and a mucilaginous ingredient such as starch or flour, in order to retain sufficient moisture to act as an electrolyte throughout the entire life of the cell. The zinc chloride, which is an intensely hygroscopic substance, aids in this also. It will be observed that this paste does not occupy

* The name "ion" is given to an atom or a group of atoms when in an electrified condition.

the whole of the available space between the zinc and the carbon, but is merely coated on the zinc pot to a depth of about quarter of an inch or so. In many of the best cells this zinc excitant, professionally known as "white lining" is a sort of extremely sticky jelly

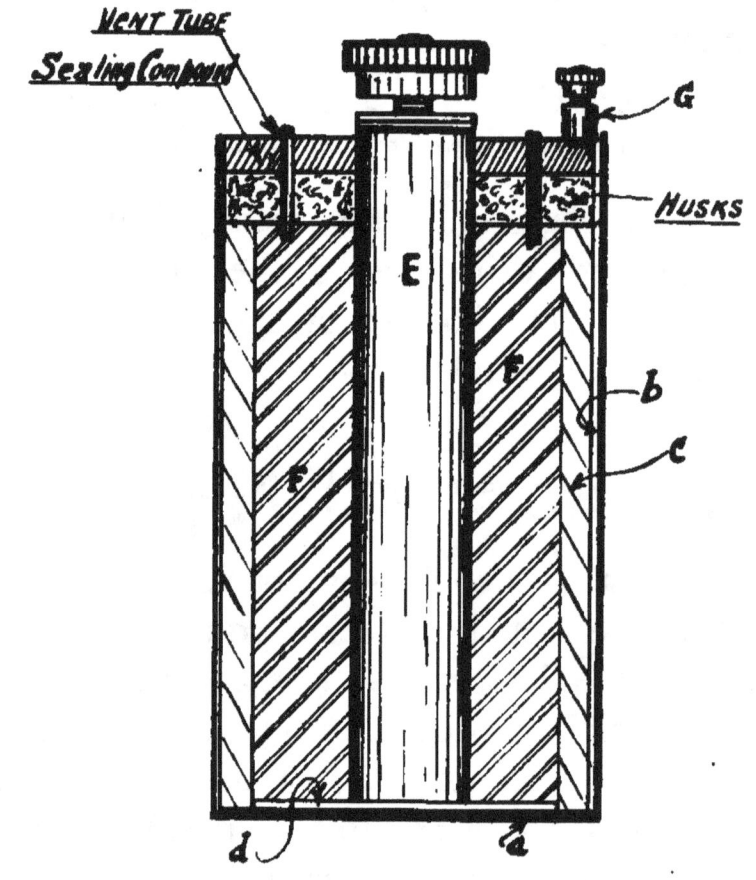

FIG. I.

which remains perpetually tacky. In others the white lining is a piece of blotting paper soaked in the electrolyte. Whichever type of zinc excitant is used it must not "dry out," or the cell will not operate. Climatic and storage conditions are most important considerations in dry-cell longevity, indeed, dry cells should be stored in the upright position in a *cool*, dry place to keep fit for active service ; (*d*) is a wad of insulating material for the carbon and its mixture to stand on.   If this wad were absent the carbon would short-circuit the zinc can, and so effectually prevent the generation of the external electro-motive force (E.M.F.) used for our ignition coil ; (E) is the carbon plate or rod itself, complete with its terminal ; whilst (F) is the excitant and depolariser, which, of course, is principally sal-ammoniac and manganese dioxide ; (G) is the connection to the zinc can ; sometimes this takes the form of a wire soldered on to the zinc, or more usually the far superior plan of fixing a brass terminal is adopted.

To complete the cell it is sealed in with black or red sealing compound, usually a wax resin mixture which will not attract moisture and will prove an effective insulation for the cell.   In Europe it is customary to leave vent holes as indicated in the figure, and the sealing composition is not run directly on to the cell contents, but a layer of cereal husks, or even sawdust, is interposed in order that the percolation of the gases may be facilitated.   In America the practice is to seal the cell up completely leaving NO vent whatever.   The generated gases are no doubt taken care of by some chemical absorbent.   The American point of view is to keep all moisture IN the cells and all outside atmospheric conditions OUT of them.   Both plans seem equally effective, so far as the author's experience goes, although the American plan is preferable for service under distinctly adverse conditions.

In the Ford ignition battery four cells like that illustrated are joined up in series ; that is, the carbon connection of the first is joined up to the zinc terminal of the second, and so on, until the four cells are connected, this leaves the carbon of the fourth cell free, and this is to be connected to one lead of the wiring whilst the unattached zinc terminal of the first cell is connected to the other lead of the ignition circuit— usually the "grounded" or "earthed" conductor.

In some ignition batteries the individual cells (sometimes five in number) are put up together in a large case or box, and sealed in *en masse*, only the two main terminals being exposed. In assembling cells into a battery care must be exercised that the zinc cans do not touch each other, for if this occurs some of the cells will be cut out and the battery would be useless, and would spoil speedily. Always keep the dry battery away from the heat of the engine, and endeavour to fix it so that the cells are upright in position. It will then last much longer. The best makes of dry cells can be kept in stock a number of years without deterioration if stored as suggested. The author has a battery of Hellesen dry cells that have been in storage for six and a half years, and the terminal voltage is still up to normal. They are of the usual circular type.

Tests on the Dry Battery. — The electrical capacity of a good dry cell is very high when compared with the storage battery, weight for weight. For example, a good dry cell, weight 1 lb. 12 ozs., upon test gave a capacity of 40 ampère hours. A storage cell of the same weight would not have anything like that capacity. If of the lead/acid type it would need to be some 7½ lbs. in weight. At first sight the reader may gather that it would be a paying proposition to use dry cells for car lighting. This, however, is not the case for many reasons, the following being the most prominent ones : (a) the inability of a dry cell to maintain either a constant voltage or amperage for any length of time; (b) the impossibility of re-charging a dry battery; (c) the high cost of running dry cells when more than an occasional current is required from them. Any one of these points is sufficient to prohibit their extensive use for the automobile lights. To illustrate what can be done in the way of lighting from a dry battery the author submits the following data of tests he has made with Hellesen's batteries. *Test* 1 was on a battery, weight 4 lbs. 10 ozs., voltage 4·5, amperage on a short circuit 15 amps. capacity in ampère hours, laboratory test maximum 40. This battery on road service furnished the current for a tungsten miniature lamp of the usual flashlight or " daylo " pattern for *two hours a week* for *nine months*. Upon calculation of the amperage gotten on the road, and contrasting this with that obtained in the laboratory, it was found that the battery had a working capacity of some 24 ampère hours. *Test* 2, made upon a larger battery of the same type, gave a more efficient performance as the *difference between the laboratory capacity and the practical service capacity* was ten ampère hours only, which compares favourably with the 16 ampère hours in the previous example. The point that the writer wishes to bring out, is the fact that the capacity of a dry battery increases at a very much greater rate than the price, therefore it is economy to employ the largest size of battery that can be conveniently carried. In ignition practice there is a saving of $33\frac{1}{3}$ per cent. by using a larger cell, although the fact that one regular size of cell, $6'' \times 2\frac{1}{2}''$ diameter, is turned out by the million has modified this calculation to some extent, in so far as the United States is concerned, and in that country a far greater saving is obtained by the use of two regular five-cell batteries installed with a change-over switch, so that one can be used one day and the other the next. This will take full advantage of their recuperative powers; indeed, it is the way to make two batteries last as long as three used in the ordinary continuous manner.

In wiring up such batteries, have the ground or earth connection as near to the engine as possible, as this may save occasional trouble through a " high resistance " or a " poor contact " to chassis. In cases as on the Ford where the dry battery is used for easy " starts " only and comparatively infrequently at that, there is no advantage in installing two batteries. The advantage is only apparent when the battery is used frequently or for lighting in any form.

The laboratory test of a dry cell will in a large measure indicate its practical value, since a cell which in the laboratory shows a capacity of, say, 40 ampère hours with the usual laboratory rate of discharge ($\frac{1}{4}$ of an ampère) is to be preferred to one having the same capacity on an $\frac{1}{8}$ ampère discharge only; indeed the chances are that the former cell would "spark" an engine better and for a longer period under working conditions. Although it must be admitted that, unlike the storage cell, there is no *one best rate* of discharge for the dry cell—a rate of working that if adhered to will allow of the utmost value for money being obtained out of any cell. This is largely due to the fact that with the dry cell we have a high recuperatory factor and a low open circuit discharge coefficient to reckon with.

The voltage of the dry cell should be carefully measured with the ordinary small testing instrument carried in the tool box (as a rule). Test each cell separately if possible and see that the contact is good against the terminals ; this is most important, as sometimes an increase in pressure or a scrape with a knife or file will send the reading up one or two points. The reading on the volt-meter should be about 1·5 for each cell ; in the case of a combined battery where it is impossible to get at each cell the reading should equal 1·5 times the number of cells which comprise the battery. Thus a four-cell battery would show a voltage of $4 \times 1·5 = 6$ volts.

As the cell gets " used up " the voltage will fall to

about 0·25. No cell is of much use for continuous ignition or lighting work when the reading is less than 1·25 volts. The above test should be made with the cells out of commission or the information gleaned from the volt-meter will be misleading. With the storage cell, on the other hand, the reverse is the case, and a light load is essential to a true indication of the cell's condition.

The amperage of the cell or battery should now be taken, *i.e.* the number of ampères it will give on "short-circuit." The ordinary cell-testing ammeter, which ranges up to 30 ampères, is the proper instrument to employ for this test. The leads from the meter should be applied to the battery across each terminal in the obvious way, but do not leave the wires on for more than the smallest period of time— just sufficient to get a steady reading on the dial. With the ordinary cell testing "dead beat" meter this can be accomplished in about half a second.

It will be evident that under these conditions the cell or battery as the case may be, in producing an amount of current limited only by its internal resistance which is of course the resistance inside the cell between all the surface of the zinc (inside) and the total carbon surface, see *b* and E on Fig. 1. The resistance of the leads and ammeter may be considered as nil. This being the case it would injure any battery to take this test more than occasionally. A word of warning must be issued here against the use of this test upon a storage battery. Such an experiment is dangerous, as the internal resistance of a secondary cell is so low that over a hundred ampères would pass out, which would be sure to damage the meter, even if it did not injure the operator's hands (by burns not shock) and the battery.

Returning to the legitimate use of the short-circuit test, it will be necessary to reject all cells that do not yield a current of about 25–30 amps. if of American make and 15–20 if of European origin at time of sale. A reduction is usually made on all "low test" cells. In some countries a price-amperage schedule has been formulated based on a rule which will shortly be expounded. When a battery which when new showed 25 ampères on short-circuit falls down to 10 ampères at the end of several months of service, one can do a worse thing than replace it with a fresh one, as the ignition is likely to miss any time after this status is reached, whilst for lighting purposes such a battery would be indeed a broken reed.

It is a fair guide to assume the following theorem in dry-cell work, although it is sadly wanting in mathematical accuracy, or any pretensions to a scientific foundation : The fraction which has for its denominator the number denoting the short-circuit amperage when the battery was first installed, and for its numerator the present short-circuit reading is equal to the amount of capacity still forthcoming [barring accidents.] For example, suppose the automobilist tests his new battery and finds a reading of 30 ampères ; after the lapse of three months he trys it again and gets a reading of 20 ampères on his meter. Applying our rule, we got $\frac{20}{30} = \frac{2}{3}$ capacity still available, and we conclude that one-third has been consumed.

Cleanliness.—The dry battery works best when clean and free from dust and dirt. It should be periodically inspected and dusted. Have no oil cans, spanners, monkey wrenches or tyre levers near it so that they could become jolted across the terminals of the battery.

Fixing.—Fix the battery at the back of the car as far away from the heat of the engine as possible. It should be installed in such a manner that it cannot jolt or fall over on its side. It is best to have the battery fixed with its terminals on top, not lying on the side on end or upside down. By observing this little point the battery will last longer in service.

Re-charging Dry Cells.—To attempt to resuscitate dry cells by passing a charging current through them, is a popular fallacy and waste of time and energy—electrical and otherwise. It is true that certain types of dry cells that have " gone dry," can be given a new lease of life by drilling through the sealing compound and soaking them in a half saturated solution of sal-ammoniac, but the process is tedious and at the best uncertain, and it is rarely worth the trouble. These remarks do not apply to the various European cells of the export type which appear to be sent out bone dry and require the addition of water just as the storage cell requires acid before it will work. These special dry cells are constructed with an internal irrigation system which carries the added moisture to every part of the cell, so that in the course of a few hours the cell absorbs it just as black loam will absorb water. It has been the author's experience with these cells that there is an element of uncertainty about the powers, and those he used in a South African expedition failed to produce more than two ampères on short-circuit at any time of their existence.

CHAPTER II

THE LEAD STORAGE BATTERY

WE are now to delve into the inner mysteries of the "STORAGE BATTERY" of the U.S.A., or the "ACCUMULATOR," as it is familiarly called in Europe. In scientific literature it is often spoken of as the "Secondary battery" to distinguish it from the cells of the type dealt with in chapter one—*primary cells*. The distinction between the primary and secondary cell is more arbitrary than real. It is quite possible that at any time some modification of the well-known primary cells, such as the Leclanché, Bunsen, Bichromate, Daniell, etc., might largely enhance their practical reversibility, and thus render them commercially useful as " storage " cells.

Exception has been taken to the terms " Storage Battery" and "Accumulator" on the premise that such cells neither store nor accumulate *anything*. This is pushing matters to the extreme ; for, although the popular notion that such devices *store electricity !* AS ELECTRICITY, is not true, there is a very real storage of energy—chemical potential energy—resulting from the conversion of the electrodes or plates, and the electrolyte from a low to a highly energised state. This in some measure justifies the employment of the familiar nomenclature. Furthermore, as it is electricity that produces these highly energised states, and since electricity results from their undoing, cells do, in effect, if not in fact, act as retainers of electricity. Theoretically, there is no difference between the primary cell and the secondary, since they are both apparatus for the conversion of chemical energy into electrical energy. In reality every secondary cell is a primary cell, and every primary (including the dry cell) a possible secondary cell.

The superiority or otherwise of a storage cell depends simply upon how near its component electrodes approach to the ideal of reversibility, which is reached when one has produced an accumulator in which the passage of an electric current in the opposite direction will exactly reverse the chemical changes that took place whilst discharging, and restore the cell both chemically and physically with the minimum expenditure of electrical energy.

In spite of prolonged experimentation, until quite recently one substance alone was found to be worth while considering for commercial purposes, all other substances formed only cells that were more fitted for the museum than the market. The substance universally adopted was lead. Its real superiority results from what may be termed the *accidental* suitability of the chemical behaviour of lead, and certain of its compounds to the purpose in view—the ideal of electro-chemical reversibility as just set forth. The

C

reversibility of the lead cell is very marked, and, given considerably less weight, the ideal cell would be not far distant. Electrical regeneration can take place some hundreds of times before the slight chemical and physical changes that do take place prevent its further use. With every other form of cell (excepting the alkaline species) that has been devised, the changes that are inherent to charging and discharging are so great that the cells become unserviceable before reaching maximum capacity, and repairs and renewals are a standing order. The reason why other combinations than lead were tried will be apparent to every chemist, as the lead/lead peroxide cell offers but a poor energy-storing capacity when contrasted with other metals, weight for weight.

Essentials of a Storage Cell.—Every storage cell consists of three essential parts:—*The box*, which is the cell or vessel that holds the solution or electrolyte. *The plates*, electrodes, or elements, and *the Electrolyte* itself, in which are immersed the positive and negative plates. The plates, both positive and negative, virtually consist of two parts—the active material and the support. In the pasted or Faure type of plate the support is usually called the "grid." The active material is that portion which actually undergoes the chemical changes during the cycle of charge and discharge. It may be prepared in many ways and from numerous materials, but when completed and in its fully charged state it always consists of lead peroxide (PbO_2), the chocolate coloured oxide of lead at the positive pole, and lead (Pb) in a more or less finely divided and spongy condition at the negative electrode. The quality of the active material is most important, as the efficiency of the cell depends upon its porosity. Indeed, those portions to which the acid or electrolyte has not molecular access are useless for storage purposes. It will be observed that it is this accessibility that distinguishes the negative active material and its support, as chemically they are both lead. In the manufacture of the plates all the material which is pressed into the grid is spoken of as "active" material, although in practice a comparatively small portion of this is really "active."

The student can easily emulate the historical experiments in connection with the storage battery, indeed much can be learnt from the repetition of Planté's famous experiment, a modification of which is roughly sketched out in Fig. 2 which show two strips of sheet lead connected by insulated wires to a suitable battery, "dry" or otherwise. The solution in the jar is accumulator or battery acid, the exact specific gravity is not important in this case. The two pieces of lead should face each other as shown. When connected up with the battery, it will be observed that myriads of bubbles rise from each piece, the one connected with the negative pole of the battery in particular. Presently the piece connected to the positive terminal will turn a deep chocolate hue upon its inner surface. After this has gone on for a minute or two, disconnect the battery and place a voltmeter across the wires attached to the lead strips. It will be noticed that a current of electricity is detectable. We have in reality made a storage cell of very small capacity, since the quantity of active material present is exceedingly minute, as the lead strips are not laid out in such manner as would encourage its formation and retention, also there is only one

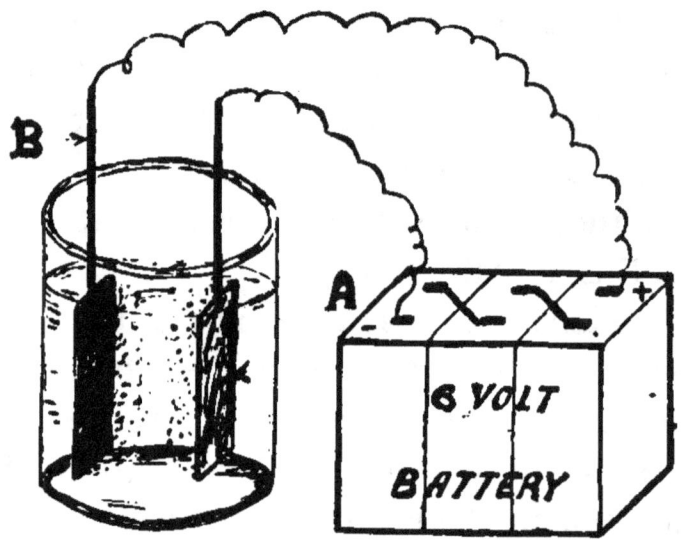

FIG. 2.

surface of each plate working. Doubtless the reader will remember that all storage batteries have in each cell an unequal number of plates. There is always an extra negative plate—the negative being cheaper than the positive. In our experiment if we were to put another lead strip opposite surface A, but not touching it, and connect it by a wire to plate marked B, we should have raised the capacity of the cell as *both* sides of plate "A" would then work. This, then, is the reason for the extra "negative" in the storage battery. To encourage the formation of active material it is necessary to charge up our strips and then discharge them, and next time upon charging to change over the battery wires so that the positive strip will now become a negative electrode and *vice versâ*. Repetition of this process would ultimately "spongify" the surface of the plates, and if we were to score their inner surfaces with a rough file we should considerably increase the storage capacity. This stage of porosity is equivalent to the "formed" state, and our cell now needs a long charge in one direction, and no further reversals, or the active material will "shed." Factory made cells are fully formed when they reach the consumer, and must NOT be charged in the wrong direction or they will be ruined. The effects of which we have spoken are the results of "electrolysis"—the decomposition of a liquid by an electric current. Extremely active substances are liberated from the acid in the cell which are able to cause the physical and chemical changes in the strips of lead. For the benefit of the chemically inclined reader, the following largely hypothetical information is usually included in all works on the subject of storage batteries :—During discharge both plates (chocolate-coloured *positive* and slate to purple grey *negative*) are slowly converted into lead sulphate ($PbSO_4$) with the removal of "sulphion" (SO_4) an as yet unisolated compound of sulphur and oxygen existing in sulphuric acid. Thus the reduction of specific gravity of the acid is accounted for. The action on the positive plate is a little more complex, and takes place in two stages—or at least is said to do so. First, the peroxide of lead—the active material—is reduced to a lower oxide, lead monoxide (PbO), and this is subsequently converted into lead sulphate. This sulphate is a perfectly healthy, normal sulphate, and is necessary to the action of the cell. The pathological "sulphate," of which so much has been written—the white plague of the storage battery—is an abnormal deposition whose origin is still stranded in darkest mystery, in fact even the joint authors of "*Occult Chemistry*" shed no light upon this perplexing problem. However, a certain brand of Storage Battery is now upon the market which is undoubtedly manufactured by people who have some inside information upon this "sulphate" problem as this particular battery is GUARANTEED not to "sulphate." Whether this "*only real improvement in the lead storage battery in twenty years,*" is due to occult qualities of the active material or to super-excellence of construction the makers, in their modesty, omit to state.

During charging the reactions of discharge occur in reverse order and a situation like the one illustrated in the following equation is not unlikely :—

POSITIVE PLATE.*

$$PbSO_4 \quad + O \quad + H_2O \quad = PbO_2 \quad + H_2SO_4$$

(Normal lead sulphate) (Nascent Oxygen) (Water from the solution) (Lead peroxide) (Sulphuric Acid given back to the solution from the plate)

* Actually these plates are incorrectly described. Compare with plates in primary cells.

NEGATIVE PLATE.*

$$PbSO_4 \qquad + 2H \qquad = H_2SO_4 \qquad + Pb.$$

(Normal lead   (Nascent Hydrogen)   (Sulphuric Acid from   (Lead)
sulphate)   the plate)

Present day storage batteries are often built up
with plates equivalent to our lead strips and are

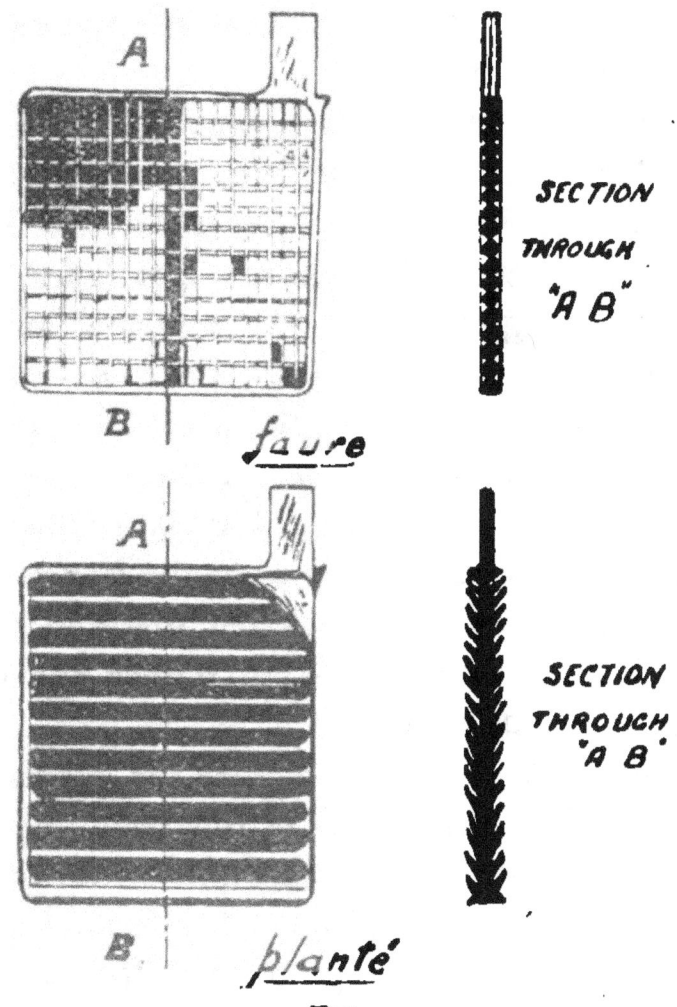

SECTION
THROUGH
"A B"

faure

SECTION
THROUGH
"A B"

planté

FIG. 3.

known as Planté type cells. A rough sketch of a
Planté or formed plate is given in Fig. 3, lower

* See note on p. 22.

portion. From the views shown it will be gathered that plain sheet lead is not employed, but a plate is cast having shelf-like abutments on each side upon which the active material sits. It does not merely rest on these ridges, but most firmly adheres thereon. The forming process usually takes about a week to complete, and is carried out in much the same manner as in our experiment.

It will be observed that the upper plate shown in Fig. 3 is of a different order from the Planté. The construction is typical of the most widely used plate, the Faure type, patented in 1881. The holes, as will be apparent by glancing at the sectional view, are tapered upon the inside, not so much as shown in the sketch, but merely enough to retain the active material. The great difference between these two plates is that in the Planté type the active material is produced electrolytically, whereas in the Faure pattern the plate is pasted with raw active material to help along the forming process and so reduce the cost of formation to a minimum.

Efficiency Considerations.—As the generation of electricity by the storage cell is the result of certain chemical reactions which take place between the active material and the electrolyte or acid used in the cell, it is of no advantage to load up the cell with a quantity of raw material that will remain quite out of reach of the molecular action. Some cells are rendered inefficient owing to the mistaken notion that the more plates one can get into a cell and the thicker these are, the better will one's battery work. Not only does the practice prevent an adequate amount of acid being contained in the cells, but the majority of the paste is of no use whatever, since the acid can never react with it.

In the automobile starting and lighting cells and in the electric automobile, too, the battery has to be cut down to the last ounce. Many precepts and practices applicable to the ideal cell have necessarily been sacrificed, in order to construct a battery of light weight and high discharge rate. This result has been accomplished by using only active material of the highest specific output, by the employment of high gravity acid the maximum permissible (1·300), and, lastly, by reducing the support or grid portion of the electrode to the lightest possible proportions.

It must be here stated the strain upon the plates when discharging at a high rate, such as would occur when the engine is "on compression," or when an "electric" is "bucking" a grade, is enormous. Therefore it is not to be assumed that mechanical strength of plate is of secondary importance. The heat evolved during the high discharge adds its quota of destruction too. However, just as it is impossible to crush an egg by squeezing it between the palms of the hands, providing it is pressed upon its long axis, so it seems impossible to buckle a well-constructed battery plate —at least for some time.

The table taken from a manufacturer's catalogue will indicate to the sceptical that makers anticipate the current required by a refractory "self-starter."

STARTING BATTERIES—6 VOLT.

Rating in ampère hours.		Momentary Starting Rate. amps.	Dimensions. Add ¼ inch to length to determine length over handles.			Wt. (lbs.).
When discharged at ¼ of capacity per hour for 8 hours.	When discharged at $\frac{1}{10}$ of capacity per hour for 10 hours.		Length (inches).	Width (inches).	Height over all. Including handles (inches).	
80 amp. hours	85 amp. hours	250	$8\frac{1}{8}$	$7\frac{3}{16}$	$9\frac{7}{8}$	40
93 ,, ,,	100 ,, ,,	325	$9\frac{1}{4}$	$7\frac{3}{16}$	$9\frac{7}{8}$	47
,, ,, ,,	,, ,, ,,	,,	$9\frac{7}{8}$	7	$9\frac{7}{8}$	47
106 ,, ,,	115 ,, ,,	400	$10\frac{3}{4}$	$7\frac{3}{16}$	$9\frac{7}{8}$	55
125 ,, ,,	135 ,, ,,	450	$11\frac{1}{2}$	$7\frac{3}{16}$	$9\frac{7}{8}$	63

The Containing Cell or Box.—In Europe celluloid boxes were the vogue a few years ago, and even to-day the public will have them for their automobile batteries. Celluloid is a pernicious substance for an accumulator case, and possesses just three points in its favour—its lightness, its transparency, and its immunity from breakage. Most of the storage battery troubles can be traced to the use of celluloid boxes and separators—especially this latter. The acid and its nascent derivatives act upon celluloid and decompose it ultimately to a sugar-like material. Hard rubber or ebonite of a flexible, or at least semi-flexible nature, is the favoured product in the U.S. storage batteries, and its popularity is increasing in those of European manufacture. Glass cells are sometimes used, and are featured by two leading English firms. Carrying cases should be painted inside and out with acid-resisting paint. Individual cells should be separated by means of a thin partition.

The Insulators or Separators.—Prolonged usage has determined that for all portable batteries the one substance for a separator is WOOD. Hard rubber or ebonite was at first thought to be the ideal, and many electrical engineers (including the writer) used this material as a substitute for celluloid. However, as some of these cells came in for repair, the tearing down revealed that this material too was "decayed" by the electro-chemical action. As a rule, it is found that after three years of continuous service the hard rubber separator will crumble to dust like an Egyptian mummy exposed to the atmosphere. Perforated or corrugated wooden separators are treated with a solution which neutralises their natural acid; in some cases they are treated with paraffin wax, boiled in it, in fact, so that the electrolyte will not affect them. Woven glass is used in the well-known Van Raden accumulator and appears to be eminently satisfactory. Rubber bands is another idea which the author has seen. This is unfortunately open to the same objection as ebonite.

In designing a separator the chief points to keep in mind are (*a*) ample separation, (*b*) non-impedence to passage of acid and gases in process of charge and discharge, (*c*) provision for the prevention of internal short-circuits through falling active material and bridging the space between the plates in any way. In automobile work a separator the whole width of the plate is essential.

Terminals and Connectors.—The importance of terminals and connectors can hardly be over-estimated, as, nowadays, when we expect a current of some 400 ampères (see above table), a little misfit or loose connection in the wiring from the electric self-starter to battery, no matter how trifling, will render impossible such a draught of current (see Ohm's Law) as our pushing force is never more than 12, and in most cases only 6 volts. Using Ohm's Law we can ascertain that the internal resistance of the battery plus the resistance of the cables attached to its terminals must be less than 0·015 of an ohm! To accomplish such a result it is necessary to have the most positive contact at the battery. The old pattern finger-nut terminals are inadequate unless the leads have a good cable socket attached to them, and the terminal nuts are tightened with pliers. A connection block fitted to the side of the battery box is a better plan, the leads being sweated into a tapered brass cap end, which in turn fits tightly into a female socket on the connection block, and is held there by two set screws, which have their points so constructed that they bite into the metal of the cable tag.

Another plan very popular in America is to have the cable tag of hard lead tapered and screwed at the upper extremity, so that it can be jammed into the lead receptacle at the top of the battery lug.   When this has been done a few threads are accessible at the opposite side of the lug and a lead terminal nut is screwed thereon thus ensuring a perfect non-loosening contact, which needs tapping with a wooden mallet to loosen it when disconnection is desired.   Always see that the terminals, and, in fact, the whole box, are clean and free from acid and dirt.   A little vaseline is useful for keeping the battery in proper condition.   It should be applied with a rag, special attention being paid to the lugs, terminals, and metal carrying handles.

CHAPTER III

CHARGING STORAGE BATTERIES

BEFORE commencing to recharge any battery, it is necessary to ascertain that it is in a fit condition to receive a charge, as much labour and expense can be saved if a little attention be given a battery before the current is run through it.

The point of first importance is to see that there is sufficient electrolyte in the cells—examine *every* cell. The level of the acid should be kept a quarter of an inch above the tops of the plates. To ascertain if this is the case, proceed as illustrated in Fig. 4. The tube is an ordinary piece of glass tubing $\frac{3}{16}''$ bore such as is used in the old (insanitary) type feeding-bottles for babies, it is inserted down the vent hole where the acid is poured in until it strikes the top of the plates. Place the thumb over the top of the glass tube as shown on the right of Fig. 4, and withdraw the tube carefully. The height of the liquid left in the tube will indicate the amount by which the plates of the battery are covered. If no liquid is left in the tube, try it again to make sure, because if the thumb does not shut out the air properly, the liquid will drop back into the cell.   In a large battery the acid level might be ⅝″ above the plates with safety.   If level is too high it will seep out over the tops of the cells. The test having failed, fill up to ¼″ level with *pure* water (*not* tap or faucet water).   If there is good reason to believe that acid has been spilt through accident, add acid of proper specific gravity, which in America is 1·285 to 1·300, in Europe an acid of

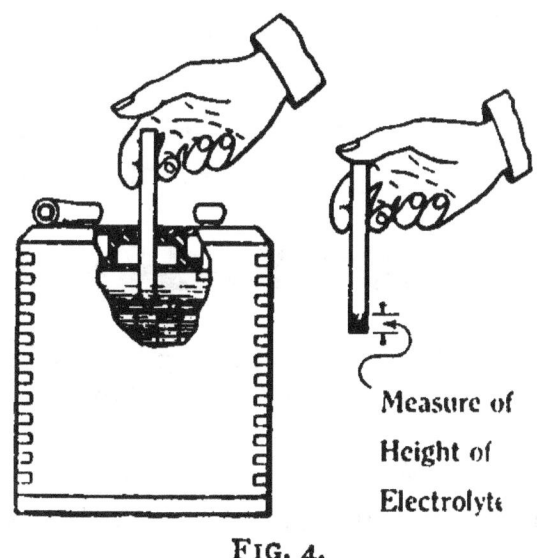

Measure of Height of Electrolyte

FIG. 4.

lower gravity is recommended 1·175–1·200. ·C. A. Vandervell of England uses an acid of 1·225, which corresponds to the discharge gravity of the American types.   Personally the author always uses an acid of 1·225, as he thinks that the higher figures are too near the limit of safety.   If the gravity is above 1·300 the risk of injurious sulphation is greater, and the ampère-hour efficiency is lower than with weaker electrolyte ; furthermore, if sulphation does set in, it is more difficult and sometimes impossible to remove, whereas, if the lower density acid is employed its removal by prolonged charging in a weak electrolyte is *comparatively* easy.

For testing the acid there are two very popular forms of apparatus, one is known as the Syringe-Hydrometer, which consists of a fair size glass, rubber-bulbed syringe, with a rubber tube to insert down the vent-hole of battery, the body of the syringe is large enough to contain sufficient liquid to float a small glass hydrometer, which indicates the specific gravity of the liquid drawn up from the cell. The other form is that shown in **Fig. 5**, and does not require so much

FIG. 5.—Yellow bead, 1·170 s.g. discharged ; blue bead, 1·190 s.g., fair condition ; purple bead, 1·225, s.g., fully charged battery.

intelligence to use as the former ; it is called the Bead Hydrometer, and consists of a white-backed glass tube with a rubber bulb at one end, and at the opposite end is fitted with a small rubber tube, like the syringe type. Three coloured and carefully weighted beads are inserted in the tube. When the acid solution is drawn up into the tube, the beads rise when the various gravities are reached. The readings are taken when the beads reach the centre of the tube, and will denote respectively Battery Discharged, Fair Condition, and Fully Charged, as they rise separately in the order stated in Fig. 5.

The method of using the bead type of hydrometer is depicted in Fig. 6.

It must be noted that when a small quantity of acid or water is added to a cell, it is better to test the gravity of that cell after charging has gone on for at least 15 minutes, in order to give the acid a chance

FIG. 6.—The above illustration shows the method of using the hydrometer, which operates in exactly the same manner as an ordinary fountain pen filler.

to mix in thoroughly. The reading may be too low or too high if this precaution be neglected.

How to make Battery Acid.—In making up electrolyte for the battery, always add acid to the water, and NEVER *vice versâ*. Pour the strong acid into the water in a steady thin stream, and have the mixing vessel of earthenware surrounded by water if possible, as heat is generated in this operation. The water,

D

as elsewhere stated, must be pure distilled. The acid must be pure *yellow brimstone* sulphuric of 1·843, this is purer and better suited for battery work than the kind manufactured from iron pyrites. The volumetric proportions are, strong acid one part, water four parts; do not test the s.g. until the mixture is cool. It is advisable in many cases where accumulators can only be charged at infrequent intervals, to add one part by volume of the " anti-sulphating " solution to the following strength of acid, which is of the correct specific gravity :—

Acid (A).
Distilled or (failing that) rainwater 19 parts, (yellow brimstone) acid 5 parts, soda solution B 1 part. Add the sulphuric acid slowly to the water, and finally add the soda solution, and stir well.

Soda (B).
Sodium carbonate (Na_2Co_3). Take a saturated solution of the above salt (cold), soda solution ½ pint, pure sulphuric 3 oz.

This special acid prevents to some extent the formation of the injurious " higher sulphates " of lead, providing ordinary care is used.

Some people like to use a " jelly acid " so that the battery solution cannot be spilt. It is not good practice, but for those who wish to use it, here is the formula.

Acid sulphuric, 1·250 s.g. to be mixed with a dilute solution of sodium silicate, 1·180 s.g. in the proportion of three of the acid to one of the silicate solution ; the resultant will be a colourless solution on first making, which must be poured into the accumulator as it slowly sets; after twenty-four hours the mixture assumes a pale-blue tinge, and will be a jelly.

There should be a little free acid on top of the plates, about $\frac{1}{4}$ inch or so.

The current used for charging purposes must be of the right form and of the right strength, that is to say, its flow must be uni-directional, therefore an alternating current supply will not serve, also if its density be too high, the plates will be injured.

How NOT to charge a Battery.—The author has known quite a few people who have made a direct connexion between the light supply and their car battery. When they have "switched on," generally there has been a loud report and a flash somewhere—and the fuse has "blown" (melted). This is what happens in the fortunate cases, but instances are on record where no fuse of a suitable gauge was in circuit. Possibly the fuse had "blown" at one time and, as no suitable fuse was at hand, a hair-pin was used in its stead. The heat generated by the large rush of current would cause the cell to take fire and the accumulator has presented the appearance shown in Fig. 7, reproduced by courtesy of Messrs. Richford & Co., London. Occasionally one hears of a cell burning up in the process of charging. The cause of this is a slow heating effect, and can be produced in many ways; from actual experience the author selects a few:—

(*a*) Scarcity of acid in the cell—a drop or two only perhaps in one corner of the case—and a great deal of sediment. When the current is passed through the cell the high resistance of this acidulated silt causes heating, and if the case is made of celluloid it

Hard rubber battery overloaded, and under supplied with acid.

FIG. 7.—Celluloid-cased battery burnt out by a short circuit.

may take fire. When this happens the case does not burst into flames, as one might expect, but burns slowly away with a little cloud of white smoke having the smell of camphor. To avoid this contingency, be sure to have plenty of acid in the cells before charging. Loss from evaporation must be made good from time to time or the electrolyte will get low. Attention is required every three weeks at least, and when touring in places like Arizona or the Karoo *every week* is the time for a little distilled water to be added to the battery.

(*b*) Accidental fusing of connexion wires between certain batteries joined in series. This is due to faulty terminal connexion and occurs only through gross carelessness and cannot happen with a low voltage supply. On a 240 volt circuit, for instance, it would be quite possible to have an arc start up owing to the heat which would occur if a dirty piece of connexion wire were to be used on an equally dirty terminal. Either through age or the heat the wire would snap, and immediately, full voltage is available at the fracture which starts an arc.

(*c*) The commonest cause of firing is due to leakage of acid from one or more extreme corners of the box. It is a common practice in a garage to charge up batteries on a lead-covered bench. This lead may be connected to the supply accidentally, or, in technical language, become " grounded " or " earthed." Leakage from the cell will provide the circuit, and slow combustion not unfrequently takes place. It is usually more frequent with old celluloid accumulators, as the box on account of its age and the chemical action referred to on page 26 is a good conductor of electricity. Suspicious-looking cells should always be stood on a plate or a piece of wood to insulate them from the metal bench.

There are three general methods of charging: Firstly, from primary cells; secondly, from a direct-current dynamo; and, thirdly, from the electric-light supply. The cheapest way is undoubtedly by means of a small dynamo, driven by means of water power or gas. For successful charging the dynamo must be "shunt" wound, and be fitted with a laminated cogged drum armature, wound in as large a number of sections as possible in accordance with its size. It must be fitted with adjustable brush gear and lubricators on both bearings sufficiently large to accommodate a plentiful supply of oil, as the speed will not be much under 3000 revolutions per minute. If a small water turbine to work off the cellar tap is contemplated such a plant as illustrated in Fig. 8 will give 15 volts 3 amps. on 70 lbs. of pressure, and upon 45 lbs. will give 8 volts 2 amps., which is sufficient to charge up a 6-volt car-lighting battery. Fig. 9 gives an impression of the switch board supplied for such a set, and consists of an ammeter of suitable range, also an adjustable resistance unit. Should an oil or gas engine be decided upon to furnish the motive power for the dynamo it will be best to select one of about half-horse power, as engines below this power wear out rather quickly as they border on "toyland." The author ran a small storage battery plant with a $\frac{1}{4}$ h.p.

FIG. 8.—Water turbo set.—Minimum water pressure required
45 pounds.

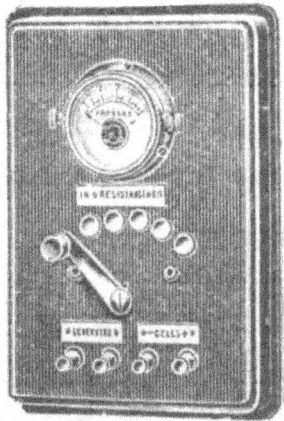

FIG. 9.—Switchboard for turbo-generator.

gas engine satisfactorily, but its fuel bill and replacement account was as much as that for an engine of twice its size and power. Fig. 10 gives an idea of the type of plant required by a small garage. The engine is of $\frac{1}{2}$ h.p. and the dynamo furnishes 8 ampères at 25 volts pressure, and will charge four of the largest automobile batteries of 6 volts each.

With this type of generator the windings are well protected, and, above all, there are few lines of force

FIG. 10.

wasted, owing to the iron path being relatively large, and completely enclosing the magnetic field, enabling more than 85 per cent. of the total magnetic flux produced to be utilised in the armature. The centre of gravity is conveniently placed, it being neither too high (as in the over-type), nor too low (as in the under-type machine). This goes a great way to ensure even and steady running. Having started up the plant, run up the speed of the dynamo until it produces a voltage equal to that of the combined cells on charge. If a 6-volt and a 12-volt battery were to be charged in series the voltage of the dynamo would need to be about 18 volts before switching the cells in circuit. The voltage of the dynamo will then adjust itself automatically to that of the back E.M.F. of the battery, but the engine speed should be increased so that the scheduled charging current is put through the cells. This information can be found on the manufacturer's name plate on battery box as a rule. If everything is O.K., connect the cells, taking care to connect them the right way round, which is the negative terminal of the accumulator to the negative of the dynamo, and the positive of accumulator to positive of dynamo.

If in doubt as to which pole is positive or negative, try the action of the wires on moistened pole-finding paper, which paper can be readily made by soaking pieces of white blotting-paper, the size of a postage-stamp, in an aqueous solution of starch and potassium iodide, and then drying.

In making the solution of iodide of potassium care should be taken not to have it too concentrated.

When a test is desired, moisten the paper with a little water and place the two bared wires whose polarity it is wished to determine on the moistened portion about $\frac{1}{8}$ inch apart. As soon as the current passes, a dark brown or blue-black stain will evidence itself at the *positive* wire. Another even simpler test for polarity is to immerse the two wires in a glass of water, which has been acidulated with a little accumulator acid in order to render it a conductor. A mass of bubbles will be apparent at both wires, but the *negative* wire will have the larger congregation.

The cells may be charged by one long charge, or by two or three short ones ; it makes very little difference, providing, of course, that they are not discharged in the meantime. " How do you tell when an accumulator is properly charged ? " is a question which has very often been asked. The usual way is to place an ordinary moving iron-type voltmeter across the terminals and charge until the reading is about 2·5 volts *per cell* in battery ; thus a 12-volt battery would read 15 volts ; but, unfortunately, the decimals of a volt vary with the differing resistances of voltmeters. For example, a good moving *coil* voltmeter will take below 0·01 of an ampère, whilst the ordinary rough-and-ready moving *iron* cell tester will take more like 0·1 of an ampère. A better test would be to put a lamp across its terminals, and note the brilliance of the light. After a little practice the user will be able to tell to a nicety whether his accumulator is charged or not. The lamp used for testing *must* be a carbon-filament one of voltage suited to battery under test ; otherwise our test will be inefficient, because the filament of the metallic lamp of any commercial value is insensitive to slight differences of pressure ; it is just this property that makes them so useful for small work.

If the voltmeter shows a good voltage *whilst the*

lamp is also connected to its terminals, a good charge can generally be depended upon ; as a discharged battery could not show a reading of correct value for many seconds under the circumstances.   There is still one other test that may be useful should the cells be transparent ; this is to note whether the acid be "milky," or, better, turbid, *after* charging has ceased for some fifteen minutes or so ; this always denotes a thorough charge with an accumulator that is in good condition.   The opaqueness is due to the myriads of bubbles present in the acid ; they can be seen with a good hand lens.

A similar and companion test to the above is to note the colour of the plates, and whether they are "gassing" freely.   When they have been doing so freely for a period of at least one hour, one may safely reckon on a thorough charge.   After a cell has been charged for some time and is still having current passed through it, it will bubble and froth, similar to the froth that is seen on alcoholic beverages. Now that this stage has been reached, the negative plates will be purplish-grey in colour, whilst the positive ones will be of a deep chocolate hue ; it is then advisable to stop charging, as the electrolyte is uselessly suffering decomposition.   When the acid is placed in an accumulator it should have a specific gravity of 1·225 ; on charging, it will gradually rise until it attains its maximum gravity, between 1·275 and 1·285, when it is "fully charged."   The above is the most reliable test, but charging should be continued until there is no further rise in the density of the electrolyte.   An ingenious person has made the above fact evident, by constructing an accumulator as shown

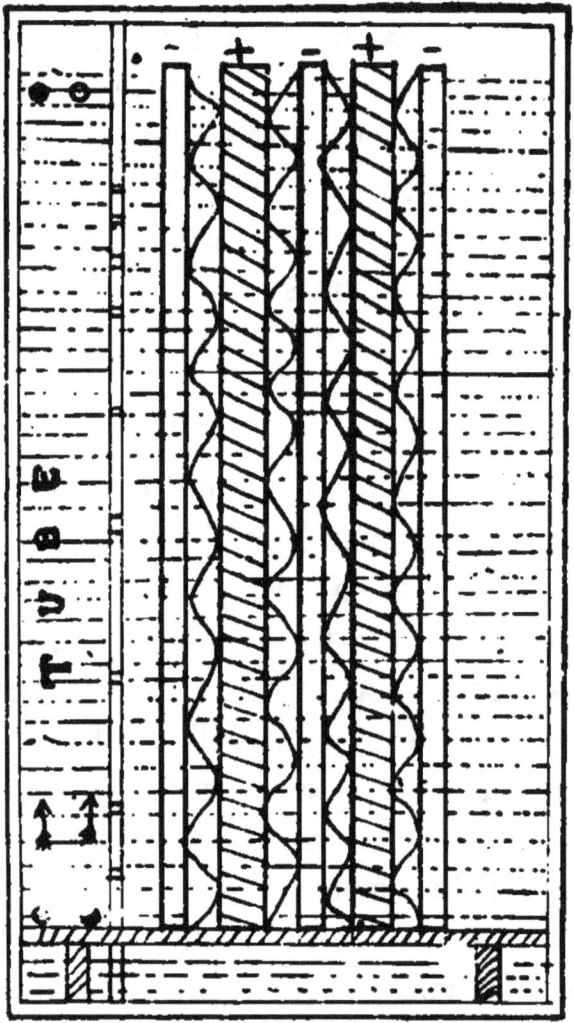

FIG. 11.

in Fig. 11.   It is an ordinary type of cell with the addition of a perforated tube, in which are placed some beads of a special composition possessing different colours and densities.   Assuming that the cell is discharged, a solitary red ball (the lightest) is floating at the top of the tube; when it is, say, half-charged, two green balls float up to the top; at the completion of the process, a blue ball rises from the bottom of the tube and rejoins its comrades at the top. By this means the state of the cell can be seen at a glance. A similar device is used in hydrometers (see Fig. 5). There is yet one great difficulty to overcome in connexion with " dynamo charging," *i.e.* the event of a *voltage drop*. In the ordinary course of things it would be a serious matter supposing the belt to come

off, or the water or gas to fail, as the back pressure of the charging-cells would reverse the state of affairs, and turn the dynamo into a motor, thereby damaging both cells

FIG. 12.

and dynamo. To overcome this impediment, an automatic cut-out switch (see Fig. 12) can be fitted with advantage ; but, for those who do not wish to go to the trouble and expense, it will be an efficient safe-guard to place an inch or two of fuse-wire in circuit, with accumulators and dynamo of such a gauge that it will just carry the maximum current required without getting unduly hot. The principle involved is that, should the dynamo fail to generate, the back-electro-motive force of the cells would tend to drive a large current through the machine, and a heavy discharge quickly ruins the dynamo and also accumulators of the usual type. This is obviated by use of the fuse or automatic cut-out. The charging rates for the various sizes are given below, and their *discharging* rate should not exceed 50 per cent. on charge rate.

Size in Ampère Hours.		Charge Rate.
9 ampère hours	. . . . .	0·5 to 1·3 ampères
18 ,, ,,	. . . . .	1 ,, 2·6 ,,
27 ,, ,,	. . . . .	2 ,, 4 ,,
45 ,, ,,	. . . . .	4 ,, 6·5 ,,
63 ,, ,,	. . . . .	7 ,, 9 ,,
90 ,, ,,	. . . . .	10 ,, 13 ,,

The above rate is for the ordinary type of lead cell. It is advisable to vary the rate of charge throughout the period. For example, in the case of the 45-ampère-hour size the *commencing* rate should be 5 ampères, whereas the *finishing* rate could with advantage be reduced to 1·25 ampères only.

Should supervision be impossible it is best to charge the battery at a 24-hour rate of 2 ampères until charged. The *finishing* rate should be in operation when the battery is gassing well and should be kept up until the s.g. of the acid has ceased to increase.

Charging from Primary Batteries.—There are many people who do not wish to run a dynamo because the first cost of itself and accompanying " motor " is great. For such persons there are two methods open—either to use the current from the house supply (if such there be) or to use more or less satisfactory batteries. We will take the latter case first. There are two distinct types of cell which are admissible: the Daniell cell or the double-fluid chromic-acid cell. The latter is the better by far, but it requires more attention to keep it in working order. It will also charge four times as quickly as a battery of Daniell cells. There are, of course, several other excellent cells of even greater merit than these, as far as output is concerned; but, unfortunately, they are more expensive, and not a few deal with mineral acids in their concentrated state, which give rise to objectionable vapours if much current is taken from them; so we had better confine our attention to the above-named cells. If we decide to do our charging by means of a battery of Daniell cells, we must note that their electromotive force is only 1 volt (1·1); hence we shall need five such cells connected in series to perform the operation. These cells, to be kept in condition, require to be connected to a high resistance * when not in use, as the Daniell cell is essentially adapted for closed circuit work. It is possible to use the gravity type of Daniell; but this is not so convenient as the ordinary porous-pot type, as the least shock causes a disturbance of the two liquids, which are separated by gravity only, and upset the voltage of the cell *pro tem.*, which would be unfortunate, to say the least, if an accumulator were connected at the time of the occurrence, as a vigorous electrolytic action would probably ensue.

To construct our battery we shall require five earthenware jars of about one quart capacity.† Each jar will require a semicircular piece of sheet copper to

---

* An 8 c.-p. carbon glow lamp such as is used on 220 volt lighting circuits will answer for a high resistance.

† 3 lb. preserve jars will answer the purpose.

form its positive element.   Also, five porous cells, as thin as can be obtained, will be required (old Leclanché porous pots will answer our purpose after the contents have been extracted).   Inside each porous pot there is to be a stout amalgamated Leclanché zinc.   To use the battery, fill the outer jars with a concentrated solution of copper sulphate (bluestone), adding a few crystals to replenish it from time to time, and also add a small quantity of strong sulphuric acid.   Into the porous cells pour a sufficient quantity of sulphuric acid (1 of acid to 12 of water) to fill them to the same level as the liquid in the outer jar.   Before using the porous cells it is as well to place them mouth downwards into melted paraffin-wax to a depth of 1 inch, as this prevents the solutions from mingling too quickly. When all is ready, connect the cells in series (zinc of one cell to copper of next, and so on), taking care to connect the positive of the 4-volt accumulator to the first copper, and the negative to the last zinc, test to see if the voltage is 5 first, however.   No resistance or fuse is necessary, as the current produced seldom exceeds 0·5 ampère at 5 volts.   To charge larger cells than 10-ampère-hour it is advisable to use the double-fluid chromic-acid battery.

For a double-fluid chromic-acid battery we shall only require three outer jars and porous cells, as the electromotive force per cell is 1·9 volts.   Nine large carbon plates will next be required, with three zincs of a special type.   These can be cast in a mould from scrap zinc.

The easiest method of making the special zinc *
from scrap, is to take a length of broom-handle and
wrap a piece of sheet iron round it so as to form a
tube, which is to be kept in its tubular shape by
winding a few turns of wire around it, and securing
the ends by any convenient method. Withdraw the
broom-handle and plug up one end with hard wood
covered with blacklead, or some other material that
will not be readily combustible. Place this improvised mould into a pail of sand to keep upright
and steady, melt up the zinc, and skim off the dross
or scum which will form on the surface of the heated
metal, then stand well back and pour in the molten
zinc until tube is full. Great care should be exercised
in transferring the zinc from the ladle to the mould,
as it has a trick of spluttering at unexpected moments
—owing to the air and moisture present in the tube.

A disk of zinc will also be required ; this had best
be made out of a piece of thick sheet zinc. If it is
found difficult to obtain disks, squares will answer
the purpose ; both, however, should be $\frac{1}{4}$ inch in
thickness.

An alternative method of producing the negative
element is to purchase Leclanché zinc rods and fit
the disks or squares to these ; beside the obvious
advantages of these, it will be well to observe that
this zinc is not cast but drawn, and is thoroughly
amalgamated in process of manufacture ; a drawn

* The reader will probably be aware that the particular type of zinc
required can be bought for very little.

E

zinc has a much better surface, and also conducts the current better than a cast one.

The size of zincs will depend entirely on the height of the containing porous pot; 8 inches by 1 inch diameter is a useful size. After having first removed the superfluous sand and filed off the burrs, drill a $\frac{3}{32}$ inch hole in top of rod, and tap to $\frac{1}{8}$ inch Whitworth (or No. 348 Machine Screw). Into this hole screw a piece of No. 12 S.W.G. (or No. 10 B. and S.) copper wire of any convenient length: this serves to connect it up to the next carbon. The foot at bottom of rod is for the purpose of standing it in mercury, the reason being that by so doing it will always be thoroughly well amalgamated.

Before using the battery it is very necessary to thoroughly amalgamate the zincs by treating them with mercury, otherwise their span of existence will be short, owing to the interaction or local action caused by the impurities present in the scrap zinc. The mercury does not extract these impurities, but it covers them up by dissolving out the zinc (pure), and bringing it to the surface of the rod, enabling it to act as though it were chemically pure. Nevertheless when the battery is not in use, always remember to lift the zincs out of the porous pot, as in course of time some of the chromic acid diffuses through the pores of the pot and this *does* attack even pure zinc; this unavoidable diffusion is one of the greatest drawbacks to this method of charging, withdrawal of zincs saves 10 per cent. of their life. For the uninitiated it may be mentioned that a *porous* pot is necessary because the electricity passes along with the solution through its walls, which phenomena could not take place if the inner vessel were glazed or otherwise rendered non-porous, and subsequently no current could be obtained from the battery. To aid the amalgamating process the following precautions should be observed : First render the surface of the zinc bright and fairly smooth by means of a file or glasspaper ; then stand it on an old plate upon which a little mercury and a little dilute sulphuric acid have been poured. With the help of a piece of linen or rag rub the mercury over the brightened surface of the zinc : it will readily adhere. If any of the mercury is spilt pick it up with a piece of amalgamated zinc, as it lays hold of the quicksilver as readily as a magnet does iron filings, though not for the same reason.

It is a good plan to soak new or dry pots in water before use ; also place the elements in their places before adding the solutions, as it is of the greatest importance that the levels of the liquids in porous pot and jar should be the *same*.

An excellent substitute for the proper battery carbons is made by using electric arc-light carbons (solid ones for preference) ; use six to eight arranged in a circle around the porous pot. In using these carbons (or any carbons for that matter) that have no proper binding screw or terminal fitted, take the greatest possible care to make a good connection to each carbon either by first copper-plating their ends and soldering on a connecting wire,* or by wrapping them very tightly round with copper wire six or seven complete turns. (N.B.—Always enamel or shellac this connexion in order to protect it from the corrosive action of the liquids employed.) Success in battery work depends principally upon cleanliness and simplicity of construction coupled with scrupulous attention to the details.

To use the battery, connect it up as shown in Fig. 13, and pour dilute sulphuric acid (1 of acid to 12 of water) into the porous cells to within 1 inch of top; also add a little mercury into each pot. The switch and fuse shown in the illustration are not indispensable, but are merely fitted for the sake of convenience; so also the ammeter. Into the outer jar pour a solution (concentrated) of chromic acid to which has been added a little strong sulphuric acid. The best quantity of sulphuric can only be determined by experience, as it is largely dependent on the purity of materials utilised. The addition of a very small quantity (half a teaspoonful) of potassium chlorate increases the efficiency of the cells. The positive terminal of the storage battery is attached to

* Battery carbons may be efficiently "coppered" by connecting them to the negative or zinc terminal of a single dry cell and submerging them to the depth of about one inch in a saturated solution of bluestone (copper sulphate). The carbon terminal of the dry cell is connected to a copper strip which is caused to dip into the copper sulphate solution. Care should be taken that actual contact is not permitted between the carbon and the copper strip. Process takes twenty minutes to complete.

the carbon pole of the charging battery, and the negative is connected to the zinc terminal as shown in Fig. 13. A battery such as the above ought to charge an ignition battery *marked* 20 ampère-hours capacity (actually 10 ampère-hours) in six hours at small cost. The current obtained from the above battery varies from 1 to 5 ampères, depending upon

FIG. 13.

its internal resistance. Three such batteries joined in series are needed to charge a 12-volt storage battery. Two batteries will successfully charge a 6-volt starting battery of moderate size, with one charge of chemicals. The zincs require renewing at the end of three charges of chemicals. To ascertain whether the solutions require replenishing, dip a small slip of white paper into the outer jar ; if the paper is stained green, the battery needs fresh solutions. If orange colour the solution is serviceable. The method of charging storage batteries by means of primary cells is, unfortunately, the most expensive and most messy of all the usual methods.

Our last method to consider is that of charging from the electric supply mains.

Charging from Direct-Current Supply Mains.— In charging from electric service mains it is always necessary to use some form of resistance partly to choke the current, or else to employ a rotary converter to reduce the high voltage of the supply to a lower voltage suitable for charging purposes. This is usually the cheaper plan, exclusive of first cost of machinery. Resistances may take several forms, but those that are generally favoured are coils of resistance wire and lamps, the most convenient and least expensive being lamps for small work, but *vice versâ* for large plants. Lamp-holders or receptacles are placed on a suitable board and connected in parallel ; then the whole arrangement is placed in circuit with the battery and supply mains. This will be made clear by referring to Fig. 14. By this means the current passing can be adjusted at will by inserting one or more lamps to suit the requirements of the cells on charge. For those who do not care to adopt this method, it may be stated that a medium-sized battery could be charged in series with a number of tungsten filament lamps by making contact through a wall switch. The reason that tungsten filament lamps are specified is because the variation or loss of candle-power would hardly be noticeable with one battery in series with

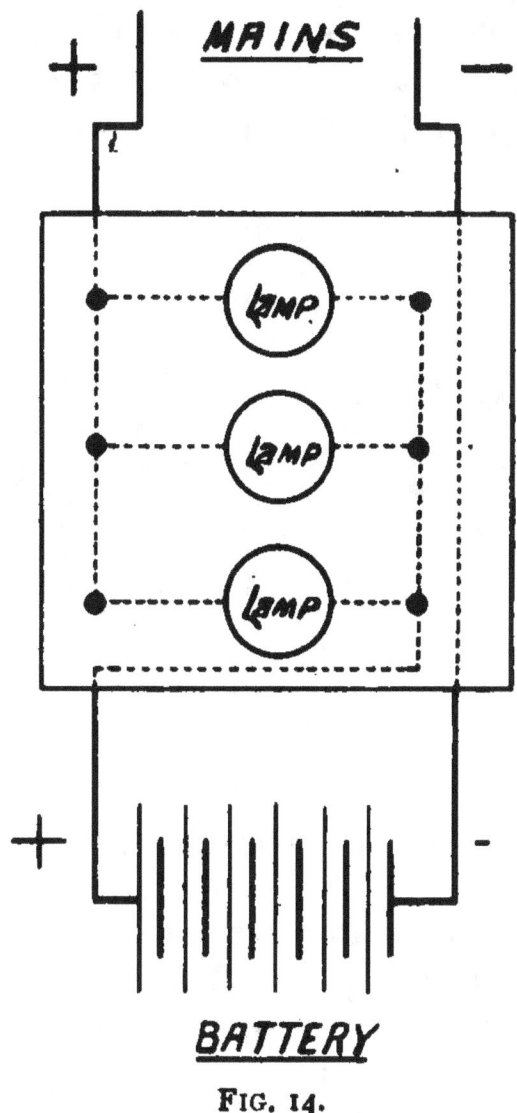

FIG. 14.

it. This method can be economically utilised in a room where a lamp or lamps are burning for several hours at a stretch. Batteries can thus be charged in

series with a lamp or lamps in the case of an elec-
trolier, whilst the lamps are giving light to the room,
and at a very small extra cost. The device illustrated in Fig. 15 shows a neat way to charge a
battery from the direct-current supply. The adapter
is put into the nearest lamp socket, and a suitable
powered lamp is inserted in the holder part. The

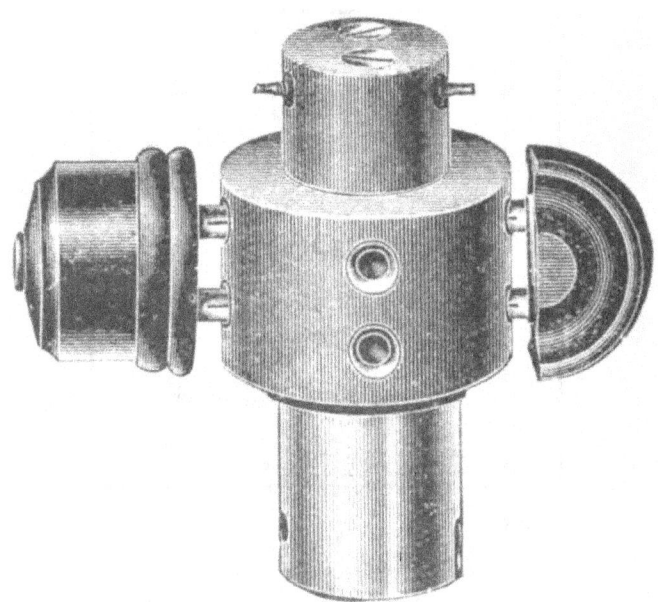

FIG. 15.—Charging Adaptor.

leads to the accumulator are taken from the two-pin
plug on the left of Fig. 15. The plug on the right is
for use in front holes when the battery is disconnected
so that the lamp may be lighted direct. Below will
be found a table giving the various currents certain
lamps will pass (approximately). It will be clearly
understood that the metallic filament lamps pass
an extremely small current in comparison with the

old-time carbon filament ones, and therefore due allowance must be made for this :—

TABLE OF CURRENTS PASSED.

Voltage.	Type.	Candle-power.	Watts.	Current in Ampères.
50–65	Carbon	8, 16, 32	30, 60, 120	$\frac{1}{2}$, 1, 2
100–110	Carbon	8, 16, 32	30, 60, 120	$\frac{1}{4}$, $\frac{1}{2}$, 1
200–210	Carbon	8, 16, 32	30, 60, 120	$\frac{1}{8}$, $\frac{1}{3}$, $\frac{2}{3}$
220–240	Carbon	8, 16, 32	30, 60, 120	$\frac{1}{8}$, $\frac{1}{4}$, $\frac{1}{2}$
50–65	Tungsten	30, 50	40, 60	$\frac{3}{4}$, $1\frac{1}{4}$
100–120	Tungsten	30, 50, 100	40, 60, 100	$\frac{1}{2}$, $\frac{3}{4}$, $1\frac{1}{4}$
200–240	Tungsten	30, 50, 100	40, 60, 100	$\frac{1}{4}$, $\frac{1}{3}$, $\frac{1}{2}$

It is very necessary to determine the polarity of the charging leads before connecting them to the battery. Use the pole-finding paper or a glass of water as directed on page 41. As in the case of charging from primary batteries always connect the positive main to the positive terminal of the storage-battery, likewise, the negative main to the negative terminal. Extreme caution must be exercised in order that the battery be not accidentally short-circuited, as owing to the construction of the ordinary lighting switches it is quite easy to do this. The battery under charge is virtually bridged across the two switch contacts normally connected by a solid piece of metal when the switch is "on." It will be seen, therefore, that if the switch is placed "on" with the battery connected across the contacts, the metal contact piece will "short" the cells. All connecting, etc., must be done with the switch in the "off" position. Shock can be avoided by taking out the lamps. If much charging in this manner is anticipated it is best to fit a two-way switch and avoid all trouble of this nature.

For those who wish to charge in a much larger way than we have been dealing with, the only satisfactory plan is to instal a motor-generator set as, where the voltage is high (220–240), the working costs of all resistances are prohibitive, and the alternative plan of "banking up" the batteries until one has sufficient to pay on a 220-volt circuit direct is inconvenient in the extreme. Therefore in such cases it pays well to put in a converter set whose primary side is suited for, say, one ampère at 230 volts and whose secondary output is about 10 amps. at 20 volts. This is a small machine for a limited scope, but any size of set can be obtained to suit the requirements of the garage or service station. Two coupled machines are not necessary although somewhat better ; a double-wound armature working within one field magnet will answer admirably.

Charging from Alternating Current.—An alternating current consists of a very rapid series of impulses, or rather waves flowing first in one direction then in the opposite direction. Such a current is of course unsuited for charging storage batteries, since the work done by one wave would be undone by the next and so on, consequently the amount "put into a battery" by the current flowing in one direction would be neutralised by the corresponding amount of current immediately following and flowing in the opposite direction.

In Fig. 16 we have the most obvious way of utilising the alternating supply for charging purposes. The small dynamo on the left of the figure is attached to the alternating current motor through a flexible coupling.   The dynamo can be wound for any

FIG. 16.—An alternate current motor coupled up by means of a flexible shaft to a small direct-current charging dynamo.

required output within the limits of the horse-power of the A.C. motor.   Thus a saving over the lamp-resistance plan used on D.C. circuits is appreciable, to say the least, although the efficiency of these small sets is usually poor, and thereby detracts much from a theoretically perfect system.   With the A.C. supply a number of other interesting systems of charging are available that are withheld from the user of the D.C. circuit, although limitations of space preclude little beyond a mere mention of the more important of these. There is the mercury-arc rectifier upon the principle of the well-known mercury vapour lamp used in most of the rapid fire photographic studios. This mercury arc has the power of blocking the inverse current and so delivers a pulsating rectified or direct current suited for charging purposes. There is one little point about these devices that is of interest to the amateur electrician—it does not matter very much which way round one connects the storage battery as the back E.M.F. of the battery is able to impress the correct polarity on the rectifier.

Another very popular type of A.C. rectifier is the Nocton Valve, a somewhat messy electrolytic contrivance consisting of jars and lead and aluminium electrodes in a solution of ammonium phosphate. Such chemical rectifiers depend for their action upon the property of certain chemical salts in solution (used in conjunction with suitable electrodes) for creating a polarising film on the electrodes when the current tends to pass one way ; this film which is formed prevents its passage and thus sorts out the alternating current as it comes by.

One of the best types of rectifier is the vibratory rectifier such as manufactured by the Wagner Electric Manufacturing Co., St. Louis, U.S.A., or that by the Adnil Electrical Co., Artillery Lane, London, England. The vibratory rectifier embodies a polarised armature which operates in the A.C. field (stray field from the transformer core). Upon switching on the A.C. supply the armature vibrates between two contacts.

Suitable adjustments are provided so that the vibrations are in harmony with the periodicity of the A.C. circuit and also the exact phase of the circuit. The contacts are closed on in such a manner that the arm connects with one contact during one half period and closes the other contact during the passage of the other half period, *i.e.* the second alternation. The apparatus is virtually a synchronised reversing or pole-changing switch. The rectified current delivered is naturally pulsating in character, as in the other devices, although, unlike that of the chemical rectifiers, both halves of the waves are passed on for use.

The great advantage of the above method of dealing with the A.C. charging plant lies in the fact that the electrical efficiency is extremely high, as the inclusion of a transformer in the A.C. circuit deals with the current more economically than any scheme of resistances in the charging circuit itself, although a small adjustable resistance unit like that shown in Fig. 9 is useful for fine adjustment.

The usual charging instructions hold good for the rectified current as for the straight direct currents. The amount of current passed will depend upon the size of the battery. It should be adjusted in accordance with the maker's instruction tag, and is generally marked upon the box. If in doubt, charge up the battery according to the schedule given on page 46.

Special Notes.—Do not discharge the battery

below 1·8 volts per cell, and when not in use always have it charged up once every six weeks. To place it in the care of a garage or service station is the best way, if the engine or plant cannot be run regularly for the hour or so required to make good the electrical loss which inevitably ensues even when the battery is out of commission.

The first charge of a battery is very important, and should be continued for 36 hours at half the rate given on the label. After the first *six charges*, empty out the acid (whilst fully charged) and refil *at once* with fresh acid of S.G. 1·225. This practice is to be commended, as it materially improves the condition of the plates by clearing away all the impurities and fragments of loose active material gathered in course of manufacture of the component parts and the assembling of them. The subsequent renewal of acid ought to take place at least every twelve months. Attention is especially drawn to the all too common error of filling a charged cell with discharged acid, *i.e.* putting acid of a low specific gravity into a fully-charged battery.

CHAPTER IV

THE REPAIR AND UPKEEP OF LEAD CELLS

THE diseases that automobile storage batteries suffer with may be divided into two classes : (1) Plate and internal troubles generally ; (2) troubles with the containing cells or boxes. Sometimes an extra consumptive-looking battery is found to be suffering in both departments. The appliances required for resuscitation and repair will be found in every "allied home," with perhaps the exception of a small mouth blow-pipe, metal-ladle, and ½-lb. soldering iron.

The solution for the repair of celluloid boxes may be obtained from all the large accessory houses, or can be made up as follows : Amyl acetate, 2 ozs. ; Acetone, 1 oz. Mix these liquids in a small bottle and add celluloid shavings, or pieces of celluloid film previously cleaned, until the solution acquires a syrupy constituency.

A little red and black anti-sulphuric acid enamel will also be required for protection of joints and for marking purposes. Furnished with the above requisites, in conjunction with a few ordinary household tools, we are now in a position to proceed with the actual work.

Repair of Box or Case.—The first case we shall deal with will be that of the usual type of celluloid accumulator, which has been carelessly put away (with more or less of its acid) in a discharged condition for several months. The above is a typical example of a common source of trouble. On examination, its case will be found to have lost most of its transparency, and to have assumed a chalky-white appearance. This is "sulphate," and not the readily soluble substance which is frequently found upon the celluloid of a new battery sent out dry and uncharged oversea, doubtless a decay product of the celluloid. The lids should now be carefully removed by means of inserting a penknife blade between the juncture of the lids and containing-case, beginning at the middle of the outside edge, and thus working round each lid until it has been completely detached. When once a start has been made, it will split away from the case without difficulty. The plates can now be lifted bodily out and examined with ease ; afterwards, let them be placed in a bucket of water.

We will leave the treatment of plates for the present, and confine our attention to cases. Should the case be glass, all that requires to be done is to wash it thoroughly with water and dry with an old dishcloth ; this also applies to ebonite cells. Most of the present-day portable batteries are made up in hard rubber or flexible ebonite boxes without partitions, and mounted in a wooden carrying case.

These boxes are sealed in with a pitch-like compound

which softens on heating with steam to such an extent that it can be readily dug or scooped out with an old spoon. The most expeditious way to steam-heat the battery, is to rig up a crude boiler made from a can fitted with a spout—an old kettle would answer well—and equipped with a piece of thin hose pipe which will pass down the vent holes of the battery. The steam is thus caused to heat up the whole top portion of the cell into which the hose is passed. After ten or fifteen minutes the pitch and the lid over the top of the plates can be removed as the steam is heating up the next cell.

In the celluloid types, too, it is more usual to use separate cells for larger batteries than 4-volt, as there is always a slight risk of the dividing partitions giving way and rendering the cells inefficient by reducing their total voltage to 2 volts, and their combined capacity to the same as that of *one* cell (since the plates in the perforated cells act only as conductors) ; also the separate-cell system admits of better insulation. Celluloid is a most convenient material to work, transparent, and extremely light, whilst glass and ebonite are somewhat difficult to work, the latter not being transparent ; both are extremely brittle, and cannot be satisfactorily repaired, but a temporary repair can be effected to a cracked hard rubber or ebonite box by means of sealing along the crack with Chatterton's or Victoria compound. The method followed is to warm the case thoroughly in the neighbourhood of the crack or leak and heat the

F

compound (sold in sticks like liquorice) by means of a match until it runs, then dab it on to the case and press well home with the fingers. If the fingers are moistened the compound will not stick to them.

With celluloid, on the other hand, a satisfactory job can always be reckoned on if care be taken. Our celluloid case must be first well washed out with warm water (*not hot*, as this will spoil the shape of the case, as it softens celluloid), and the then remaining traces of the chalky-white deposit should be scraped away by means of an old knife. After again rinsing with water, stand the case upside down to drain, and subsequently dry. When dry, pour into it a little paraffin, or coal oil, as it is called in the U.S., and scour well by means of an old toothbrush, and finally wash out the oil with warm soapy water, followed by several rinsings of cold water, and a thorough drying with an absorbent towel or cloth.

The next step is to ascertain that there is no leak between the compartments. This requires careful observation and great patience, as there must not be the slightest leak, otherwise our subsequent work will be in vain.

A somewhat easy method of ascertaining if a leak exists between the two compartments is to place a strip of bright sheet zinc in one side and a strip of bright copper or, a piece of carbon, such as an arc-light carbon, in the other. Connect a wire to the zinc, also one to the copper, and join the two wires to a galvanometer (a pocket one will answer), then fill each side with accumulator acid, and note if a deflection is observable on the galvanometer; if so a circuit exists between the two cells *vià* the partition, or, in other words, a leak is present.

In testing a 6-volt accumulator, test cells 1 and 2, then 2 and 3.

One compartment should be filled with water, and it should be carefully noted whether there is a trace of water in the unfilled side after standing. Should there be any, the place should be marked, and the water should be then emptied out and the case redried. When again dry, the marked spot should be treated as follows: Apply a clean rag to the place, after having previously steeped the rag in either benzine or petrol (gasoline); this will effectually cleanse it. Secondly, prepare the surface with a little of the celluloid solution, also the surface of a small piece of celluloid (an old photographic film, after the emulsion has been removed, will answer); the exact size and shape will entirely depend on the nature of the hole or crevice—in fact, for very small leaks, it would be quite admissible to apply a thick solution, and dispense with a patch. However, when a patch is used, apply it to the place as quickly as possible, and brush an external covering of solution over the patch, to make sure of having completely stopped the hole.

After allowing two hours or more for drying, refill with acid, and let it stand awhile. If the work be sound, fill the other side with water, and note whether there be an external leakage. If such there be, it will very probably take place in one of the two outside corners.   Its treatment will be precisely similar to the above, with the following addition : if there is any sign of a split, it is necessary to attach an angle-piece of celluloid to act as a further protection, since the wear is considerable at the corners. (Fig. 17 will make the above quite clear.)   There is still one more

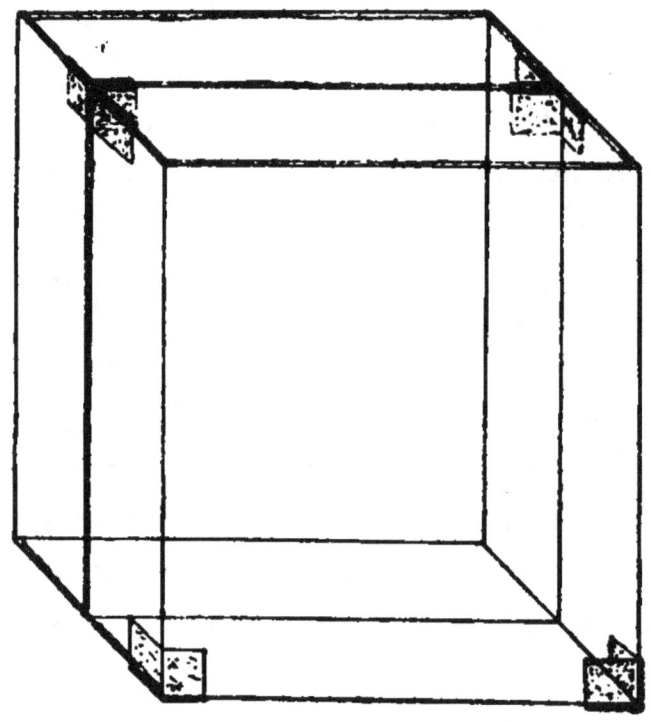

FIG. 17.

troublesome fault—*i.e.* when the partition splits away from the ends of the case.   The remedy for this complaint is to put angle-pieces *each* side of the partition, as shown in Fig. 17 by the thick lines at the junction of the partition and sides.   Fortunately, this parting (except in very badly made cases) takes place at the top, and hence is easily manipulated.   (N.B.—Before taking any accumulator to pieces, place a mark on *each* lid, and a corresponding one on each of the adjacent sides or ends, as this little precaution saves a great deal of unnecessary fitting, and consequent straining of the case.)

In bending celluloid for the corners of cases, etc., great advantage is gained by manipulating the celluloid under *hot* water, as it then becomes pliable and soft. When once bent to the correct angle or shape, immerse it in cold water to fix it thus.

Loss of Charge.—Before taking an accumulator to pieces, because of its refusal to hold a charge, examine the acid, and ascertain that its specfic gravity is correct. If found to be weak, it will in many cases be the cause ; to remedy, give the accumulator a normal charge, and then refil with fresh acid and further charge, until the correct gravity is indicated, When this state has been reached, test the cells by means of a test lamp.

Loss of charge may be due to charging the battery the wrong way round. No harm will be done the battery if the error is discovered early, but if the battery is treated thus for some time, the plates will disintegrate. Charging the correct way is the only remedy needed.

General Inspection of Plates.—*Partial short circuit* can be detected by noticing whether the positives are too light in colour and the negatives too dark. In some cases the negatives seem almost black. When the s.g. is consistently low, also the voltage or open circuit is low, a partial short-circuit may be anticipated. The cause will doubtless be sulphate or sediment at the bottom of the plates; a broken separator is also capable of causing a partial short. If the cell suspected be overcharged it will be found that the gassing is deficient in amount.

Swollen and bulging paste is usually caused by impure acid, especially if the plates are light in colour and have a blistered appearance.

Mossy Growths on top of negatives is caused by overcharging; if the plates are pitted it is a sign that level of the acid is too low. To confirm the diagnosis, test the active material of the positive plate with the point of a tooth pick; if the point enters freely and it is found to be soft at the bottom, one can be sure that the battery is being overcharged, whether mossy negatives are present or not.

Healthy Plates can be told by their colour. The negatives should have a purple to slate grey colour when charged, and a little fainter tone when discharged. The positives should always be a deep chocolate colour at the end of a charge, and of a lighter and more brownish shade at the beginning of a charge. If the positives have a fine "powdery" deposit easily brushed off with a splint of wood, and disclosing a rich chocolate underneath, the battery is getting sufficient charge. If, however, this light deposit comes away in flakes or scales, the overcharge is insufficient or not frequent enough.

Buckled Plates are due to many causes, among

which mention may be made of (1) faulty pasting causing unequal expansion through poor contact with grid; (2) charging rate too high; (3) discharge too heavy *most frequent cause*; (4) under-charging and then over-charging at too high a rate—*frequent with electric vehicle batteries.*

Sulphated Plates may be due to (1) weak acid and over-discharge, *i.e.* below 1·8 volts per cell (can be cured); (2) strong acid and over-discharge (incurable); (3) deficiency of acid (most common); (4) neglect of upkeep charge; (5) action of water or damp air on plates.

Treatment of Sulphate.—It behoves us next to turn our attention to the interior of our battery. It will be noticed that the plates are covered with a white deposit, termed "sulphate." The reader is reminded that this "sulphate" is not the ordinary lead sulphate (which appears during the usual cycle of events), but a super-normal deposition, whose exact chemical formula seems to be shrouded in mystery. We do not propose to discuss it further than to state that a sulphated plate will not perform its function satisfactorily. This insoluble sulphate cannot be readily got rid of, and the process should only be attempted when the deposition is very slight, and the paste is firm, as plates can be purchased so cheaply in these enlightened days. Assuming this deposition to be slight, and the paste firm, we proceed to remove the thickest of the deposit by means of scouring the plates (outside) with a piece of sandstone, or a piece of pumicestone will answer; both must be flat, and fairly smooth, otherwise they will take *too much* off our plates.   This scouring must be performed under water, as water is an essential, since it will carry off the superfluous sulphate as soon as the scouring frees it. As it is very awkward to introduce a piece of stone between the plates, to treat the inside surfaces, we shall find it necessary to mount a piece of either stone, in an elongated box-lid arrangement, as is employed in connexion with "oil-stones," for sharpening cutting-tools, etc.

This "stone-holder" must be of such a thickness that it will easily slide between the plates, when the stone is mounted in its position. With such a tool as is depicted in Fig. 18, the operation can be efficiently carried out.

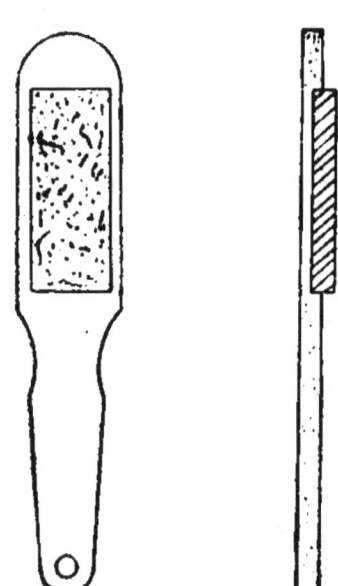

FIG. 18.

The shaded portion represents stone.

When all the sulphate has thus been removed, wash the plates thoroughly, so as to detach any adherent particles, and, finally, keep the plates perfectly dry (enshroud them in oiled paper). This will keep them fit for months—even years. In putting our accumulator together again we must be careful to cleanse the insulators or separators thoroughly, as, if any lead particles happen to be between them and the adjacent plate, it will only be courting disaster

since an internal short often results when this precaution is neglected.   Ample clearance should be
provided between the bottom of the cell and the base
of the plates in order to provide sufficient room for
any dislodged material to accumulate without in the
least interfering with the cell's action.   (When renewing
separators, use prepared wood.)   After having carefully replaced the separators in their original position—
i.e. so that they efficiently insulate each plate from
its neighbour, slide the plates into their respective
compartments, and replace the lids of each department.

Before passing on to the succeeding step, it will
not be out of place to draw the reader's attention to
a very common error, in which he is likely to partici-
pate if inexperienced in battery construction ; hence
it should be well noted that, if the best results
are to be obtained from a cell, its insulation should
be uniform, thus rendering the space between each
individual plate equal on all parts of its surface.   If
this condition is *not* complied with, it will be readily
seen that the strata of acid between any adjacent
plates will be unequal, and consequently the resistance
will also be unequal, resulting in an uneven strain on
any particular plate, and its subsequent distortion or
" buckling," as it is termed.   It might also be pointed
out that the charge and discharge rate will vary at
different points on the plate's surface, thereby causing
the above straining.

All that now remains to be done is to cement

down the lids with a little celluloid solution, applied with a brush all round the edges of the lids.

If there should be a slight gap between the edge of the case and lid, it should either be clamped together or laid on its side or end, as the case may be, and a sufficiently heavy weight applied to cause the case and lid to come together, after having previously inserted a little solution between, or, better, down the gap. Twelve hours is generally reckoned a safe time to allow for complete setting.

When set, the cells should be filled with acid and charged without delay. To reduce this sulphating tendency to a minimum, use the special acid as advised in the chapter on "Charging," *and keep the cells charged.*

If the "sulphation" be slight, prolonged charging with a low s.g. acid will in many cases suffice. At the completion of the long charge, refill the cells with fresh acid of correct s.g. Charging the battery using a solution of sodium sulphate, 200 grammes per litre of water instead of acid. A sixty-hour charge is required. Refill battery with ordinary acid.*

Renewing Plates.—It not unfrequently happens that one or more of the plates collapse, or the paste disintegrates, rendering the battery useless.

With the best makes there is a little disintegration, but that is chiefly a surface disturbance only, and will do no particular harm if cleared out periodically.

* *Journal of Physical Chemistry* described this last method of removing "sulphate."

It will be self-evident when it requires removal. Upon glancing at the bottom of the cells, it will perhaps be noticed that the deposition in question has reached the lower surface of the plates ; it then is time to empty out the acid and rinse out the powdered material with water and refill with acid ; the same acid will answer if within the twelve month limit.

Generally it is the positive section (chocolate-coloured plates) that breaks down first, owing to the fact that it bears most of the working strain. We will, therefore, turn our attention to such a case as the following :—One of the two positive plates of a 20-ampère hour battery has fallen to pieces through continual misuse or neglect. Such an example as the above requires additional attention to that mentioned in the first case, in that a new plate needs to be fitted in one of its positive sections. Obtain a plate as near like the sound one as possible, and proceed to "tin" the lug or lip of the plate as follows : With a coarse file clean all around the lug—*i.e.* remove all traces of the peroxide of lead (which composes the active materials of the positive plate) and sprinkle a little of the powdered rosin over the brightened lug ; then take up a small bead of the finest blowpipe solder on the soldering-iron and very *lightly* transfer a film of "tin" over the lug. Treat the lead casting, to which this new plate is to be joined, in a similar manner, after having taken the precaution to level off its surface so that when the new plate is affixed the whole structure will stand square, and *not* form an angle of less or more than 90°, as the case may be. This mistake can easily be avoided by cutting off sections from the casting, so that when the lug of the plate is held in contact with the prepared surface of the casting the above condition is fulfilled. To complete the operation, place the sound plate on a flat piece of board, so that the fractured casting lies uppermost. Further, rest a piece of wood of such a thickness that when the new plate is placed flat on it the juncture comes flush, or, in other words, the wood must be the thickness of the distance between the inner surfaces of the positive section in the sound cell.

The wood packing must be of a sufficient length to support the plate during the joining operation, and, further, it needs to project about ¾ inch *beyond the juncture* in the direction of the top of the casting, in order that a little plaster-of-Paris * may be conveniently placed round the juncture, forming a sort of trough, the wood being the bottom.

It will, of course, be understood that if time is of the utmost importance, strips of wood, the thickness being a little in excess of the lug, can be tacked on to the wood packing to form our rough trough or mould. All that now remains to be done is to sprinkle a little powdered rosin over the juncture, to act as flux, in conjunction with the tin, and to deftly apply a hot soldering-iron, or better, a blowpipe jet, thereby

* To be used in the form of a paste (with water) in the usual way, and left to set.

causing the metal to flow and unite. When cold, re-move all the packing, and a sound joint ought to be found ; any slight imperfection can be touched up with a well-tinned iron, finally covering the complete casting with a coat of black anti-sulphuric enamel. (N.B.—See that any enamel which flows on to the active material is at once removed, as it is very detrimental to the plate.)

Any number of plates can be treated as in the above method, both positive and negative ; but to re-new the inside plate of a section of three offers con-siderable difficulty, and a thin iron must be employed to get between the plates. Should two plates of a section of three require renewing, little difficulty will be experienced, as the *middle* plate can be put in first, without the obstruction of one above it.

It is occasionally required to renew the whole of a three or more plate-section, therefore, for a last example of plate renewing, we will take the following :—

The negative sections of a 40-ampère hour battery require complete renewal. Obtain the required number of negative plates. In the above there would be required two sections of five plates each, hence, we obtain ten negative plates the same size as the old ones, or as near as is possible, and tin their lugs. Procure a piece of well-seasoned wood (a piece of "quartering" will answer admirably) about 6 in. long, 2 in. wide and 2 in. to 3 in. deep. Bore a hole a little larger in diameter than that of the old

circular lug of the casting to which the terminal or
binding screw is attached, and of the same depth as
the leaden circular lug.   To obtain the square setting
of the casting, construct a frame of cigar-box or
similar thin wood around the hole ; the dimensions of
the enclosed space will correspond with those of the
square setting of the casting.   Fig. 19 will make this
clear.   Finish off our improvised mould by lining the

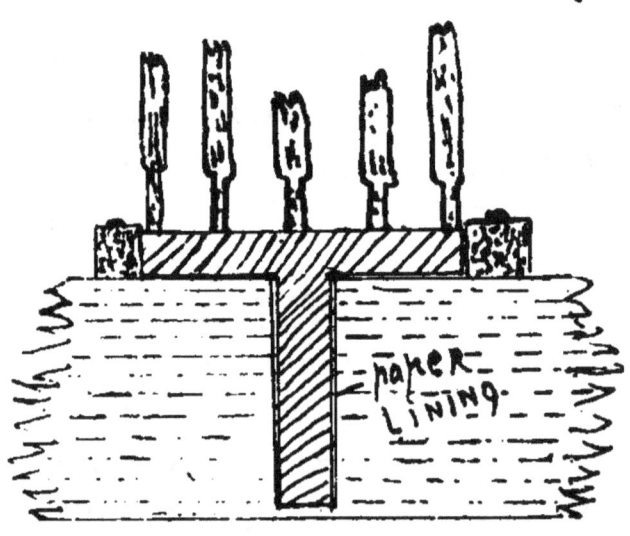

FIG. 19.

hole with brown paper glazed with French chalk or
blacklead and treat the inside of the frame similarly,
so as to obtain a perfectly smooth surface from which
the casting can be readily removed.

To complete the task we pack the plates with
wooden blocks (as we did in the first plate repaired,
four being required).   To obtain correct spacing the
blocks and plates are secured by means of several
turns of string or wire.   The greatest care must be

taken to keep the plates level and firm, otherwise we shall produce a lop-sided casting. Melt up the old broken casting with a few fresh pieces of *pure* lead, and skim the dross off the top. Hold the plates (1) in such a position that their lugs are within the square enclosure, and so that the lug of the third plate is central with the hole in the mould; (2) so that they are at right angles with the surface of the mould, or, in other words, so that they are upright. Finally, pour in the lead, and keep the plates *dead* still, until the exposed surface of the lead has assumed a " frosted " appearance; it will then have become set. Do *not* remove the work until it is quite cold, as then it can be removed with ease, owing to the prepared surface of the mould. If there should be any irregularities they can be touched up with an old file or rasp. Blow-holes, etc., can be filled in with solder. Finish off the work with black enamel, taking care not to enamel the circular lug *all* the way up, as a terminal or bridge connexion has yet to be fitted. (N.B.—It is as well to cast the circular lug longer than actually required, as then we shall have plenty of spare metal with which to produce a good square surface for fitting terminals, etc.)

Two such castings will be required for the accumulator in question. Each time the mould is used it must be carefully prepared with a little French chalk or blacklead, and the hole lined with a fresh piece of brown paper. It may be pointed out that to line the hole with brown paper can be readily done as follows:—Take a pencil or skewer about the same size as the hole and wrap a turn of the prepared paper round its circumference, and insert both into the hole. By rotating the pencil a good close fit can be obtained, and the pencil finally removed. In fitting the wood casing round the hole, take care to leave a $\frac{1}{8}$ in. space at least on each side of the hole, otherwise the circular lug will be pre-disposed to break off.

Fitting Small Terminals.—Terminals form a most important adjunct to the battery, and, we may say, a very troublesome one if allowed to corrode. Many are the devices for overcoming the corrosive action: ebonite terminals (metallic only at the contact surface), silvered terminals, and "lubricated" ones, whose containing lugs carry miniature grease-boxes. All the above have their individual advantages; but we shall only allude to the ordinary brass variety as seen on most accumulators.

To keep the terminals in a proper condition needs only a little trouble being taken in cleaning them at frequent intervals. The main cause of the deplorable condition of the average battery terminal is the practice of not troubling to wipe off any acid which may exude during the process of charging. This acid, trace though it may be, should be carefully removed with a piece of rag, and a thin coating of grease deposited on the lids of the cells; also, the terminals themselves should receive a coat of vaseline or similar mineral grease.

In this way the appearance of the battery can be kept presentable, also the energy in the cells will not expend itself in producing depositions upon the metallic terminals. To facilitate the work required in keeping them clean, dip the terminals in the thick celluloid solution, and when dry remove it off the actual contact surfaces only, thus leaving the greater part adequately protected.

There are two distinct types of small terminals in common use—viz. the telegraph pattern, in its multitude of varieties, and the winged or butterfly-nut terminal, this latter being used for heavier cables than the former one, as great pressure can be brought to bear on the contact surfaces by means of its protruding wings. When finishing off a battery, it is necessary to decide whether the terminal shall be fitted horizontally or vertically. The method of procedure is somewhat different in the two cases. In fitting a horizontal terminal, it is as well to flatten the ends of the circular lug by means of squeezing it in a vice, or by means of a hammer judiciously applied. In the centre of the flattened portion bore a hole to take the stem of terminal, but take care to have a tight fit. This stem can either be soldered so as to fix it, or secured with a nut on each side of the lug. It is usual to complete the terminal with a winged nut ; but the other pattern may be used with equal advantage. To prepare the lug for the reception of a vertical terminal, cut off the lug to the desired length, and see that the surface is smooth and

G

square. Carefully tin the prepared surface, and by means of a gimlet bore a hole in it about $\frac{3}{16}$ in. deep to take the stem of the proposed terminal after it has been well tinned. Before the terminal is inserted, sprinkle a little powdered rosin into the hole, and place a shaving or bead of solder near the bottom of its stem. All that now requires to be done is to blowpipe the *bottom* of the terminal stem by means of a small mouth blowpipe used with an ordinary wax candle. (N.B.—In blowpiping terminals, *never* allow the flame to come into contact with the lead, as being extremely hot it will cause the lead to run and melt up the lug; always let the flame play on the lower part of the stem.)

In renewing broken or damaged terminals it is quite unnecessary to remove the plates if ordinary care is exercised. Treat the lug as in the above case, after the broken metal has been extracted (use a pair of old pliers and a blowpipe flame to remove old terminals), only the greatest caution must be exercised in preventing ignition of the surrounding celluloid. It is necessary to place a piece of asbestos over, or, better, around the lug, or in default a piece of damp flannel.

Fitting Large Terminals.—In the heavy starting batteries the terminals are by no means small, and cannot be dealt with along these lines. Cables are usually fastened direct to the legs by means of all-lead bolts and nuts as referred to on page 28. Special tools are required for dealing with these heavy type connexions. To take the cells adrift it is necessary to centre punch the lug ends and drill through the connecting strips until the connector is drilled off the lug proper. When reconnecting the bars are replaced over the lugs and "burnt" in. It would be possible to do this with a candle and blow pipe in an emergency, but in practice the oxy-hydrogen blowpipe is used, as in all lead-burning processes.

Help in case of Accident.—*Acid in eye.*—If acid is splashed into the eye immediately flood eye with water and no harm will result. If irritation will not subside bathe eye with a zinc and rose-water lotion in eye-glass.

Acid drunk in mistake for water.—If acid is drunk by mistake take a drink of soap suds or baking soda in a glass of water.

Acid on clothing.—Neutralise with dilute ammonia. The red stains which appear on blue and black cloth will disappear just like magic with a dab of weak ammonia.

Acid spilt in large quantity.—Neutralise with whiting or powdered chalk.

Caution.—Do not inspect cells with match or other open flame, as the gas given off from all types of storage battery is inflammable and explosive.

CHAPTER V

THE ALKALINE STORAGE BATTERY

THE lead cell has one serious rival—the alkaline nickel iron cell. This alkaline cell is radically different in principle, material and design from any other storage cell. In the first place it contains no acid, and no lead plates are used in its construction. Instead a solution of caustic potash [or soda, a war economy!] in water is used for the liquid, and nickel hydrate and iron oxide for the active materials. For the "grid" a steel frame is used. As there is no sulphuric acid there can be no sulphation, and hence it can be left idle and discharged without sinister results.

In order to make the reader more familiar with the principles of the new storage cell, let us use the experiment performed in Chapter II., but instead of using strips of lead let us procure two very thin strips of bright steel and place them out-of-doors for a few weeks. They become "rusted." The action of the oxygen in conjunction with the moisture of the air has attacked the outer layer of the steel and formed with it an oxide of iron known as "rust."

Now let us place these two pieces of steel in a solution of caustic potash (or soda), and connect them to our battery, as shown in Fig. 2. It will be observed that one of our plates is presently changed back to metallic iron but the other is being rusted twice as much—this is the positive plate. If we connect a voltmeter or galvanometer we shall find that a current is produced which quickly disappears, and the excess of oxygen on the one plate passes back to the other, and thus the cell is "discharged." We have now charged and discharged a primitive alkaline storage cell. If instead of using two steel plates only we had used a number in each group, as with the lead cell, we should have found that the capacity of the cell was proportionately increased.

If instead of two thin plates of steel we were to construct two perforated steel pockets, and put into them a quantity of iron rust in a finely divided state we should find that upon placing these into our alkaline electrolyte, and passing the charging current as before, all the rust in one of the pockets would be converted into metallic iron, because the oxygen from the rust, which is Iron Oxide—a compound of iron and oxygen—would have passed over to the iron rust in the other pocket causing it to possess twice as much oxygen as before. This super-oxidised pocket tends to swell into a cylinder shape owing to the increase in volume. When the wires from the pockets are later connected to the instrument it will be found that an enormous increase of current and capacity is in evidence. As a matter of fact, it would be found that even though a hundred pairs of steel plates had been used, they would not have surpassed the results obtained in the experiment with the pockets. The reason for this is, that the small particles in the pockets present a much greater combined surface of active material to the solution.

We have now completed the first stage in the development of the alkaline storage battery of the present day.

Countless experiments were performed after this stage had been reached by the two great pioneers in alkaline electric storage—Thomas A. Edison (the inventor of the well-known Edison Storage Battery) and Waldemar Jungner (a Swedish scientist who first produced the Alklun-Accumulator, afterwards rechristened the Jungner-*Nife*, the " Ni " standing for nickel and the " Fe " for iron). Both these inventors found that steel by itself was not the acme of perfection, and after a wide range of experimentation with practically all the commercially possible elements, they both came to the conclusion that steel and nickel formed the most suitable combination. Like many other signal inventions, the honour of priority is a matter of dispute.

The Edison Storage Battery.—In the Edison type of alkaline cell the positive electrode consists of a series of perforated steel tubes reinforced and equidistantly spaced by seamless steel rings. These tubes are filled under pressure of some 2000 lbs. to the square inch, with alternate layers of green nickel hydrate and metallic flakes or scales of nickel. By this means the nickel is made to contact with the hydrate of nickel intimately, and thus the resistance of the active material is reduced. The completed automobile battery plate is shown in Fig. 20.

The negative plate is composed of a series of perforated steel pockets filled with rust (oxide of iron Fe_2O_3). In order to increase the electrical conductivity of the oxide a little oxide of mercury is mixed

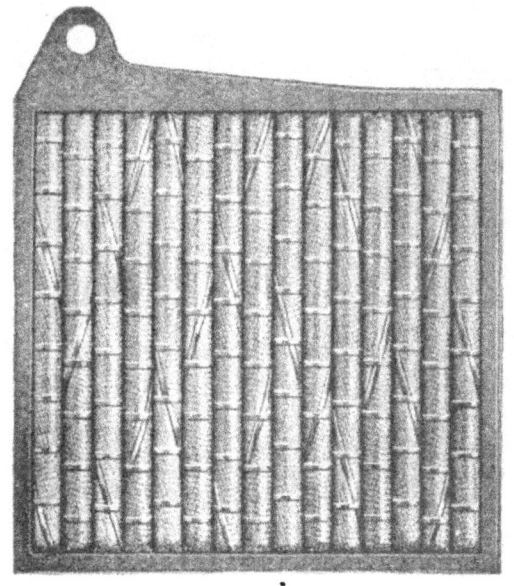

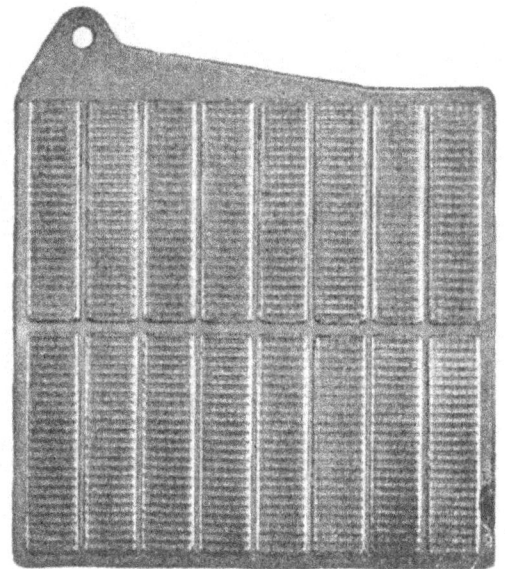

FIG. 20.—Edison positive plate.

FIG. 21.—Edison negative plate.

with it, and the whole subjected to high pressure. These pockets are then forced into the frame or grid and corrugated at the sides in order to provide sufficient elasticity for them to adhere to the oxide mass within the pockets. Fig. 21 depicts the finished negative plate to pair with the positive just referred to.

The positive and negative plates are mounted as in the lead battery, except they are a little closer together; the electrolyte, however, is NOT acid but a solution of caustic potash 20 per cent. strength, or of s.g. 1·20. As stated elsewhere, caustic soda will answer in lieu of the potash ; however, the latter is preferable.

The Jungner-Nife Battery.—In the Jungner-Nife form of alkaline cell one grid does for both

FIG. 22.—Jungner-Nife plate construction positive and negative.

positive and negative electrodes as the mechanical construction is identical, the chemicals alone differ· Fig. 22 illustrates the Jungner electrode. It may be noted that this inactive container or grid is manufactured from two very thin bands of sheet nickel finely perforated. These bands are of an indefinite length, and run one above the other through two separate rolling mills which shape them into certain profiles, the lower representing the bottom and the upper the cover. The bottom strip is automatically filled whilst running with little blocks of active material, which have, of course, been previously

FIG. 23.—Plates ready for steel can or case.

prepared. The cover and bottom are next passed through a rolling mill where they are closed up and carefully notched together along with their sides, thus forming a flat tube filled with active material. The finished tubes are then mounted together in proportionate numbers according to the size of plate required by the battery. Finally a contact frame is fitted to complete the electrode. Fig. 23 illustrates the complete automobile unit ready for insertion into its steel can or case. It will be observed that in this cell there are an immense number of plates with very little clearance between those of opposite polarity. It is quite safe to do this, as the electrolyte is to all intents and purposes inactive, and virtually acts as a conductor only. In so far as the chemicals are concerned the two cells are almost alike ; however, there are slight differences in the admixtures to the active material. For examples, whilst Edison uses flake nickel to increase the conductivity of the nickel hydrate of the positive plate, Jungner employs graphite (a kind of blacklead). In both cases these substances cause no alteration in the chemical action of the cell save to increase it. In the negative plate the conductivity of the oxide of iron is increased by the addition of mercury oxide in the Edison, while in the Jungner a little cadmium is alloyed with the iron. The electrolyte is the same in both types. The insulators or separators are in both cases hard rubber or ebonite.

General Notes on Alkaline Batteries.—*Container Notes.*—The plates of the nickel/iron cell, assembled as in Fig. 23, are placed in a steel container and welded in so that to remove them is impossible. These cans or boxes are unbreakable, and are a marked advance upon hard rubber, ebonite, glass, etc. The only " looking after " they need is to keep their exteriors free from dirt, moisture, etc., or a current leakage will result which will set up destructive electrolytes and "eat" a hole in the can eventually. A special alkali-resistant paint is supplied by the Edison people for the purpose of keeping the cells and boxes in good condition. Before placing the cells in their carrying case, always make sure that there is no dampness at the bottom.

Terminals.—The terminals and connectors of the cells do not embody lead burning in any form, but are made with stout metal clamps. A clamp connector fits over the top of a taper terminal post of steel and is forced on by means of nuts and washers. Edison cells are supplied with special "terminal jacks" for the purpose of making and unmaking connexions. As with the lead cell, it is most important to keep the terminals clean and dry. If they are left moist electro-depositions will grow upon them and set up leakage circuits, as the lids of the cells are steel and not an insulating substance as in other cells. The terminals are, of course, most carefully insulated by means of hard rubber or ebonite washers.

The Electrolyte.—The solution used in alkaline cells should be a 20 per cent. (or s.g. 1·20) of *pure* potassium hydroxide (caustic potash) in pure water; sodium hydroxide will answer almost as well. However, if any other substance is used the cells will be injured. Acids of all kinds should be avoided, and any vessel which has been used for acid should be most carefully cleansed before using it for refilling alkaline batteries. Particularly does this apply to funnels. To illustrate how difficult it is for some people to be quite sure what to put inside their battery, the author cannot do better than mention one incident he witnessed in Johannesburg, South Africa. Outside Von-Brandis Square a motorist had his auto anchored against the side-walk. Upon approaching him it was found that he was extremely busy ramming sal-ammoniac into his Edison battery with the aid of a lead pencil. This happy motorist did not bother to dissolve the salt—he just pushed it down in its crystalline condition !

Advantages of Alkaline Cells.—(a) They contain no acid, therefore give off no corrosive fumes, so can be kept anywhere without destroying surrounding objects, including the battery attendants.

(b) They can be recharged at any time regardless of the amount of charge left in, and they can be left standing idle and uncharged indefinitely without injurious results.

(c) Lighter and vastly more durable than the lead cell, double the capacity weight for weight.

(d) Can be discharged to exhaustion without injury ; indeed, a few short-circuits will not harm them.

(e) Five years of good service can be depended upon.

(f) No harm will be done to a cell if it be accidentally charged the wrong way round ; however, it will *not* store when thus treated.

(*g*) No specific gravity readings need be taken, and solution requires renewal only once a year.

(*h*) Within the limits of reason they cannot be hurt by overcharging.

(*i*) Do not run down to the same extent as the lead cell or open circuit.

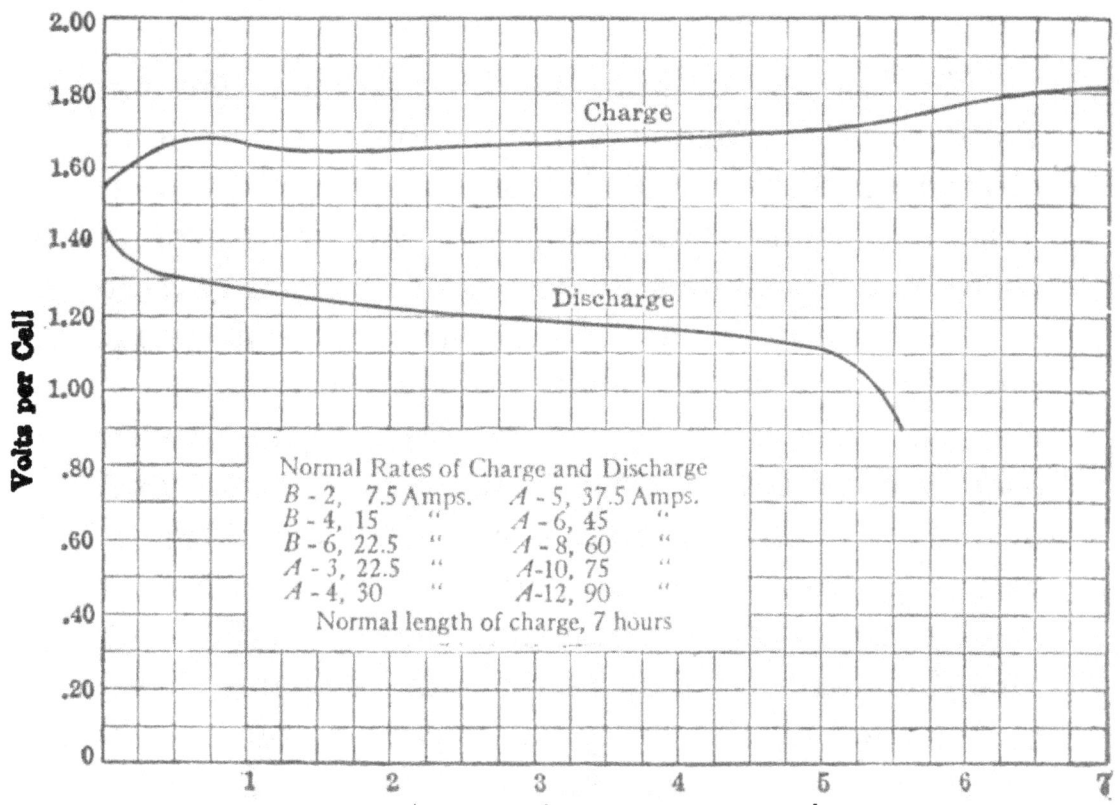

FIG 24.—Characteristics of Edison alkaline cell.

Disadvantages of Alkaline Cells.—(*a*) Voltage is very low; only 1·2 per cell can be counted on; see curve in Fig. 24.

(*b*) Internal resistance is too high for employment with electric self-starters in many cases. For all other purposes they are ideal.

(c) Initial cost is high; however, the service period being much longer in a measure offsets this fact.

The Chemistry of the Alkaline Cell.—For the benefit of the student, the author submits the following data obtained from the Edison Company. The equations have, however, been simplified for the benefit of beginners.

The fundamental principle of the Alkaline Storage Battery is the oxidation and reduction of metals in an electrolyte which neither combines with nor dissolves either the metals or their oxides. Also, an electrolyte which, notwithstanding its decomposition by the action of the battery, is immediately re-formed in equal quantity, and is, therefore, a practically constant element without change of density or conductivity over long periods of time. Therefore, only a small quantity of such electrolyte is necessary, permitting a very close proximity of the plates. Furthermore, it is unnecessary to take hydrometer readings until about three hundred cycles of charge and discharge have been made; this is simply to determine when it is necessary to empty out the old solution and put in new. The active materials of the electrodes being insoluble in the electrolyte, no chemical deterioration takes place therefrom.

Probably the simplest way of dealing with the chemical reactions is to consider that the current, or charge, in passing from the positive to the negative plate, decomposes the KOH (Potassium Hydroxide) into the ions* K (Potassium) and OH (Hydroxyl). The K ion passing with the current and carrying a charge. On reaching the negative plate the charge is given up to it, after which the atom of K unites with the water to form KOH and H (Hydrogen) is liberated as a gas thus—

$$2H_2O + 2K = 2KOH + H_2$$
(Water)

The H liberated then acts on the negative plate to reduce the rust to metallic iron, thus—

$$Fe_2O_3 + 6H = 2Fe + 3H_2O \text{ (negative)}$$
(Rust or oxide   (Iron)
of iron)

The ion OH formed on charge passes to the positive plate, and on reaching it gives up its charge to it, afterwards uniting with the water to form H_2O and O (Oxygen) is liberated thus—

$$2OH + H_2O = 2H_2O + O$$

The O liberated then acts on the positive plate in the following manner :—

$$Ni(OH)_2 + O = NiO_2 + H_2O \text{ (positive)}$$
(Nickel Hydrate)   (Nickel Oxide)

The discharge reactions would be somewhat upon these lines :—

On discharge, K goes to the positive plate, liberating H, which acts on it thus—

$$6NiO_2 + 8H = Ni_3O_4 + 4H_2O \text{ (positive)}$$
(A lower oxide
of nickel)

* An atom or a group of atoms *plus an electric charge* constitutes an "*ion.*"

OH goes to the negative plate, and O is liberated, which acts on it thus—

$$3Fe + 4O = Fe_8O_4 \text{ (negative)}$$
(Magnetic oxide
of iron)

On subsequent charge, the H liberated at the negative plate acts on it thus—

$$Fe_8O_4 + 8H = 3Fe + 4H_2O \text{ (negative)}$$

And the O liberated at the positive plate acts on it thus—

$$2Ni_8O_4 + 4O = 6NiO_2 \text{ (positive)}$$

There are probably other reactions, but those outlined above are sufficient to indicate the principles of the chemistry of the battery.

APPENDIX OF TECHNICAL INFORMATION

Volt = Unit of electrical pressure.

Compare with head of water pounds per sq. inch in domestic supply.

Ampère = Unit of current volume or rate of electrical flow.

Compare with "gallons per minute" in waterflow.

Ohm = Unit of electrical resistance — electrical friction.

Compare with friction in the water pipe.

One Volt is the difference of potential or electric level, or difference of pressure between the terminals of a "Dry Cell" (1·5 volts actually).

One Ampère is the rate of flow of current through a conductor of one ohm's resistance under a pressure or driving force of one volt.

One Ohm is the resistance or friction of 300 yards of copper telegraph wire.

OHM'S LAW

The current (I) flowing in any circuit is equal to the Electromotive Force (E) (voltage) divided by the Resistance (R) (Ohms) or friction of the circuit. Algebraically it is expressed as under—

$$I = \frac{E}{R}$$

H

To find the third component of a circuit when any two of the others are known, just cover up the one wanted and the remaining combination will indicate how to find the required factor. Care must be taken to write the formulæ as shown in equation, or the oracle won't work.

The voltage of a circuit is found by multiplying the current in ampères by the resistance in ohms. The current is found by dividing the resistance into the voltage. The resistance is found by dividing the current into the voltage.

TECHNICAL TERMS IN ALPHABETICAL ORDER

Accumulator, synonym of strange battery.

Ampère hour, one ampère for one hour, or its equivalent, *e.g.* 10 ampères for 6 minutes.

Battery, a combination of two or more cells.

Cell, a simple galvanic or voltaic couple, *e.g.* a strip of copper and zinc in weak acid is generally the unit of a battery.

Closed Circuit, condition existing when the current has a path provided for it to flow in.

Dead-beat meter, a needle movement which by virtue of controlling springs instantly comes to rest, *i.e.* does not oscillate.

Electrodes, plates of a cell.

Electrolysis, the decomposition of a liquid or moist substance by electricity.

Internal Resistance, the opposition to the passage of the current through the inside of a cell—the resistance of the path between the plates under the solution.

Open Circuit, converse of closed circuit, path broken.

" Short " or short circuit, path of no resistance or friction.

INDEX

C. A. S.

PRINTED BY WILLIAM CLOWES AND SONS, LIMITED, LONDON AND BECCLES.

USEFUL HANDBOOKS.

Electrical Ignition for Internal Combustion Engines. By **M. A. Codd.** SECOND EDITION, 120 Illus., viii. + 164 pp., cr. 8vo, 5s. net. Postage, 6d. ; abroad, 8d.

Induction—Principles of Electric Flow—Batteries—Switches—Coils—Auto Tremblers—Lodge Ignition—Distributors—Magneto Ignition—High Tension Magnetos—Faults and Remedies—Magneto Repairs—Induction Coil Design—Index.

Low Voltage Electric Lighting with the Storage Battery. Specially applicable to Country Houses, Farms, Small Settlements, Launches, Yachts, etc. By **Norman H. Schneider.** 23 Illus., 85 pp., cr. 8vo (S. & C. SERIES, No. 26), 2s. net. Postage, 3d.

Introduction—The Storage Battery—Estimating the Installation—The Electric Plant—Some Typical Plants—Installation and Operation.

The Grouping of Electric Cells. By **W. F. Dunton.** 4 Illus., 50 pp., fcap. 8vo, 2s. net. Postage, 3d.

Dry Batteries, how to make and use them. By a Dry Battery Expert. With additional notes and 30 original illustrations by **Norman H. Schneider.** 30 Illus., 57 pp., cr. 8vo (S. & C. SERIES, No. 7), 2s. net. Postage, 3d.

Modern Primary Batteries, their Construction, Use, and Maintenance. Including Batteries for Telephones, Telegraphs, Motors, Electric Lights, Induction Coils, and for all Experimental Work. By **Norman H. Schneider.** 54 Illus., 94 pp., cr. 8vo (S. & C. SERIES, No. 1), 2s. net. Postage, 3d.

Electrical Instruments and Testing. By **Norman H. Schneider.** Including Testing Telegraph Wires and Cables and Locating Faults by **Jesse Hargrave,** Assistant Electrical Engineer, Postal Telegraph Cable Company. Fourth ed., 133 Illus., xxiv. + 256 pp., cr. 8vo, 5s. net.

Laws of Electricity—Galvanometers—Rheostats—Voltmeters and Ammeters—Wheatstone Bridge—Portable Testing Sets—Testing Resistance—Potentiometers—Calibration Curves—Condensers—Telephone and Telegraph Cable Testing—Generator and Wiring Testing—Testing Telephone Lines—Switchboard Tests—Locating Grounds—Insulation Test—Measurement for Crosses—Fault Locating—Index.

E. & F. N. SPON, LTD., 57, HAYMARKET, LONDON, S.W. 1

USEFUL HANDBOOKS.

Practical Electrical Engineering for Elementary Students. An elementary laboratory course for students of electrical engineering in Trade and Technical Schools. By **W. S Ibbetson,** B.Sc., Assoc.M.Inst.E.E. 61 Illus., 155 pp., cr. 8vo, 5s. net. Postage, 4d. ; abroad, 6d.

The experimental proofs of the fundamental proofs of Electrical Engineering described in this book are divided into six groups :

(1) Those dealing with Current Measurement.
(2) The Measurement of Resistance.
(3) Fall of Potential and its applications.
(4) Potentiometer Measurements.
(5) Photometric and Power Measurements.
(6) Motor and Dynamo Principles.

In every experiment the Author describes the Apparatus required the Principle to be proved, the Method of Working, and the Method of Tabulating and Plotting the Results.

Practical Electrics. A Universal Handybook on everyday Electrical Matters, including Connexions, Alarms, Batteries, Bells, Carbons, Induction, Intensity, and Resistance Coils, etc., etc. Ninth ed., reprinted from the First. 124 Illus., 135 pp., cr. 8vo (S. & C. SERIES, No. 13), 2s. net. Postage, 3d.

The Study of Electricity for Beginners. Comprising the Elements of Electricity and Magnetism as applied to Dynamos, Motors, Wiring, and to all branches of Electrical Work. By **Norman H. Schneider.** 54 Illus. and 6 tables, 88 pp., 8vo (S. & C. SERIES, No. 6), 2s. net. Postage, 3d.

Induction Coils and Coil Making. A treatise on the Construction and Working of Shock, Medical and Spark Coils. By **F. C. Allsop.** Second ed., new imp., 125 Illus., xii. + 172 pp., cr. 8vo, 3s. 6d. net. Postage, 4d. ; abroad, 6d.

Induction—Hints on the Construction of Coils generally—Shock and Medical Coils—Accessory Appliances for, and the Application of Medical Coils—Spark Coils—Experiments with Spark Coils—Batteries for Coil Working—Faults in Medical and Spark Coils—Figures produced by Electric Discharges on Photographic Plates—X-ray Photography—Index.

E. & F. N. SPON, LTD., 57, HAYMARKET, LONDON, S.W. 1

USEFUL HANDBOOKS.

Induction Coils; how to make, use, and repair them. By **Norman H. Schneider** (H. T. Norrie). Second ed., 79 Illus., 285 pp., cr. 8vo, 5s. net.

Coil Construction—Contact Breakers—Insulations and Cements—Condensers—Experiments—Spectrum Analysis—Currents in Vacuo—Rotating Effects—Gas Lighting—Batteries for Coils—Storage or Secondary Cells—Tesla and Hertz Effects—"Roentgen" Rays and Radiography—Wireless Telegraphy—Index—Bibliography.

Electrical Circuits and Diagrams, illustrated and explained. Comprising Alarms, Annunciators, Automobiles, Bells, Dynamos, Gas Lighting, Motors, Storage Batteries, Street Railways, Telephone, Telegraph, Wireless Telegraphy, Wiring and Testing. By **Norman H. Schneider.** Cr. 8vo.

Part I. Second ed., 217 Illus., 72 pp. (S. & C. SERIES, No. 3), 2s. net. Postage, 3d.

Part II. Alternating current generators and motors, Single-phase and Polyphase transformers, Alternating current and Direct current Motor Starters and Reversers, Arc Generators and circuits, Switch, Wiring, Storage battery diagrams, Meter Connexions, etc. 73 pp. (S. & C. SERIES, No. 4), 2s. net. Postage, 3d.

The Petrol Engine. A Text-book dealing with the Principles of Design and Construction, with a special chapter on the Two-Stroke Engine. By **Francis John Kean,** B.Sc. (Lond.), First Class Honoursman in Engineering: Formerly Head of the Motor Car Engineering Department of the Polytechnic School of Engineering, Regent Street, London, W., etc. Second ed., 126 pp., 72 Illus., demy 8vo, 6s. net. Postage, 4d.; abroad, 7d.

General Principles—Description of a Typical Petrol Engine—Engine Details—The Valve—The Carburettor and Carburation—Ignition and Ignition Devices—Lubrication—Cooling—The Points of a Good Engine—Two-stroke Engines—Horse Power and the Indicator Diagram—Liquid Fuels—Appendix I.—Engine Troubles—Timing the Ignition—Appendix II.—Index.

The Diseases of Electrical Machinery. By **Ernst Schulz.** Edited, with a Preface, by **Silvanus P. Thompson.** 41 Illus., 84 pp., cr. 8vo (S. & C. SERIES, No. 38), 2s. net. Postage, 3d.

Continuous Current Machines—Single-phase and Polyphase Generators—Single-phase and Polyphase Induction Motors—Transformers—Examples of Efficiency Calculations.

E. & F. N. SPON, LTD., 57, HAYMARKET, LONDON, S.W. 1

www.ingramcontent.com/pod-product-compliance
Lightning Source LLC
Chambersburg PA
CBHW080839220526
45467CB00008B/2326